Building Digital Experience Platforms

A Guide to Developing Next-Generation Enterprise Applications

Shailesh Kumar Shivakumar
Sourabhh Sethii

Building Digital Experience Platforms: A Guide to Developing Next-Generation Enterprise Applications

Shailesh Kumar Shivakumar
Bangalore, Karnataka, India

Sourabhh Sethii
Jammu, Jammu and Kashmir, India

ISBN-13 (pbk): 978-1-4842-4302-2
https://doi.org/10.1007/978-1-4842-4303-9

ISBN-13 (electronic): 978-1-4842-4303-9

Library of Congress Control Number: 2019931830

Copyright © 2019 by Shailesh Kumar Shivakumar, Sourabhh Sethii

This work is subject to copyright. All rights are reserved by the Publisher, whether the whole or part of the material is concerned, specifically the rights of translation, reprinting, reuse of illustrations, recitation, broadcasting, reproduction on microfilms or in any other physical way, and transmission or information storage and retrieval, electronic adaptation, computer software, or by similar or dissimilar methodology now known or hereafter developed.

Trademarked names, logos, and images may appear in this book. Rather than use a trademark symbol with every occurrence of a trademarked name, logo, or image we use the names, logos, and images only in an editorial fashion and to the benefit of the trademark owner, with no intention of infringement of the trademark.

The use in this publication of trade names, trademarks, service marks, and similar terms, even if they are not identified as such, is not to be taken as an expression of opinion as to whether or not they are subject to proprietary rights.

While the advice and information in this book are believed to be true and accurate at the date of publication, neither the authors nor the editors nor the publisher can accept any legal responsibility for any errors or omissions that may be made. The publisher makes no warranty, express or implied, with respect to the material contained herein.

> Managing Director, Apress Media LLC: Welmoed Spahr
> Acquisitions Editor: Shiva Ramachandran
> Development Editor: Laura Berendson
> Coordinating Editor: Rita Fernando

Cover designed by eStudioCalamar

Cover image designed by Freepik (www.freepik.com)

Distributed to the book trade worldwide by Springer Science+Business Media New York, 233 Spring Street, 6th Floor, New York, NY 10013. Phone 1-800-SPRINGER, fax (201) 348-4505, e-mail orders-ny@springer-sbm.com, or visit www.springeronline.com. Apress Media, LLC is a California LLC and the sole member (owner) is Springer Science + Business Media Finance Inc (SSBM Finance Inc). SSBM Finance Inc is a Delaware corporation.

For information on translations, please e-mail rights@apress.com, or visit www.apress.com/rights-permissions.

Apress titles may be purchased in bulk for academic, corporate, or promotional use. eBook versions and licenses are also available for most titles. For more information, reference our Print and eBook Bulk Sales web page at www.apress.com/bulk-sales.

Any source code or other supplementary material referenced by the author in this book is available to readers on GitHub via the book's product page, located at www.apress.com/9781484243022. For more detailed information, please visit www.apress.com/source-code.

Printed on acid-free paper

I dedicate this book to . . .

My parents, Shivakumara Setty V and Anasuya T M who blessed me with their love and strength, My wife, Chaitra Prabhudeva and my son Shishir who shared their time and support, My in-laws, Prabhudeva T M and Krishnaveni B who provided help and courage, and to all my school teachers to bestow lots of love and knowledge.

—Shailesh Kumar Shivakumar

To lovers of books . . .
I hope this book aids you to get your work out to a wider audience. I would love nothing more than to see the book in the hands of people everywhere—students in the classroom, researchers, browsers in the bookstore, and professionals.
That's a great challenge, but it is certainly worth an attempt.

—Sourabhh Sethii

Table of Contents

About the Authors xvii

About the Technical Reviewers xix

Acknowledgments xxi

Introduction xxiii

Part I: Requirements and Design 1

Chapter 1: Introduction to Digital Experience Platforms 3

    Boundaryless Banking Enabled by Digital Technologies 4

    Overview of DXP 4

        Key Tenets of a DXP 5

        DXP Reference Architecture 5

        Evolution and Drivers for DXP 11

    Overview of Banking Experience Platform 16

        Key Tenets of Banking Experience Platform 16

        High-Level Requirements of Banking Experience Platform 17

        Three Ps of BXP 21

        Sample Technical Capabilities of Banking Experience Platform 21

        Sample Key Performance Indicators of Banking Experience Platform 24

    Digital Imperatives for Modern Banks 25

    Summary 26

Chapter 2: Gathering Requirements 27

    Functional Requirements 32

    Experience Requirements 36

        Seamless Experience on All Supported devices 37

        Seamless Experience on All Supported Browsers 38

TABLE OF CONTENTS

- Multilingual Requirements 38
- Navigation Elements, Menus, and Search 39
- Mobility Requirements 41
- Nonfunctional Requirements 43
- Scalability Requirements 44
- Performance—Response Time, Throughput, Utilization, Static Volumetric 46
  - Performance Requirements 46
  - Page Hits Analysis 48
- Maintenance Requirements 50
- Versioning 52
- Rollout 52
- Security Requirements 53
- Disaster Recovery Requirements 57
- Accessibility Consideration 58
- Chapter Summary 59

Chapter 3: Design 61
- Building an Experience Platform 61
- Digital Platform Strategy 65
- Platform Design Phases 69
- Design of Various Layers 70
- Presentation Layer 72
  - Scripting Framework 74
  - UI Management 75
  - UI Deployment 76
- Integration Layer 77
  - Loosely Coupled Integration and Highly Coupled Integration 78
- Business Layer 84
- Data Layer 86
- Middleware Layer 87
- Social and Collaboration Design 89

TABLE OF CONTENTS

IoT Integration Design 93
    IoT Case Study 95

Blockchain Design 96
    What is Blockchain? 96
    What Is a Distributed Ledger? 97
    Smart Contract 97
    Blockchain Platforms 98
    DXP and Blockchain Network 98
    Blockchain Components 99
    Blockchain Case Study 100

Big Data and NoSQL Design 102
    Big Data and NoSQL Integration 102
    Big Data and NoSQL Case Study 105

AI Automation Design 106
    Determine Automation Goals 106
    Steps to Build AI Automation Model 106
    Chatbot Case Study 107

Enterprise Search Engine 109

Augmented – Virtual Reality Integration 111
    Presentation Layer 111
    Integration Service Layer 112

Recent Trends in DevOps 113
    Containerization 113
    DevOps – Continuous Integration (CI), Continuous Deployment (CD) 114

Chapter Summary 115

Part II: Development of the Banking Experience Platform 117

Chapter 4: User Interface Design 119

Key Features 119
    Simplified Approach 119
    Intuitive Architecture 120

TABLE OF CONTENTS

    Dashboard 120

    Responsive Interface 120

    Personalization 121

    Internationalization and Localization 122

    Preferences 122

    Integrated Analytics 122

    Search Engine Optimization 123

User Interface Components 123

    Pages 123

    Layouts 123

    Navigational Router or Navigation Menu 124

    Presentation Component 125

    Design Goals 125

    Communication Between Presentation Components 126

    Hooks 127

Development Process 127

Development Life Cycle 129

Architecture 130

DXP UI Technology Stack 132

Angular Technology Stack 133

Angular Core 134

    Angular Support Library 135

React Technology Stack 137

    React 137

    React Support Library 137

Evaluating UI frameworks 139

    Data Flow 139

    Language 139

    Performance 139

Best Practice 140

BXP – Case Study 141

TABLE OF CONTENTS

    Consistency Across Locations 141

    Consistency Across Application 141

    Unified and Collaborative Approach 142

    BXP UI Layouts/Containers 142

    BXP Dashboard 142

  Chapter Summary 147

Chapter 5: Designing the Integration Layer 149

  Integration Consideration 150

  Data Formats 153

  Integration Services 155

  Integration Styles, Protocols, Systems, and Patterns 157

    Integration Styles 157

    Integration Protocols 158

    Integration Systems 161

    Integration Patterns 162

  Data Standards 164

  Flexible Integration Middleware 165

    EAI vs. SOA vs. ESB vs. Microservices 165

    Mutual Memorandum of Understanding (MOU) 167

    Service Protocol and Data Format 167

    API Management 167

  Why Do We Need Data Transformation Capabilities in DXP? 167

  Integration Technology Stack and Architecture 168

    Monolithic 168

    Microservices 170

  ESB and API Gateway 170

  Integration Security 171

    Authentication and Authorization 171

    Protocols 171

    Frameworks 171

TABLE OF CONTENTS

Integration Best Practices 173
BXP Case Study 176
    Case Study Conclusion 179
Chapter Summary 179

Part III: Securing the Banking Experience Platform 181

Chapter 6: DXP Security 183

DXP Security Framework 183
    DXP Layer-Wise Security 184
Common Security Scenarios of DXP 187
    Password Standards 187
    Session Management 188
    Information Management 188
    Data Validation 189
    Service Security Management 189
Security Vulnerabilities and Best Practices of DXP 190
Security Testing Framework for DXP 192
    Secure Code Scanning 193
    General Web Security testing 194
    Application-Specific Security Analysis 195
    Threat Profiling of Transaction Management in Banking DXP 195
    Threat profiling of Fund Management in Banking DXP 196
DXP Security Checklists 196
    DXP Architecture and Design Phases Security Checklist 196
    DXP Information Management Security Checklist 197
    DXP Authentication and Session Management Checklist 197
    DXP Network Communication Management Security Checklist 198
    DXP Input Validation Security Checklist 198
    DXP Security Auditing and Logging Security Checklist 199
Chapter Summary 199

TABLE OF CONTENTS

Chapter 7: DXP Information Security 201
Information Security in DXP Solutions 201
Implementing Defense in Depth 202
Firewalls and Proxies 202
Server Hardware Level Protection 202
Monitoring Infrastructure 202
Backup Jobs and Synch Jobs 203
Disaster Recovery and Business Continuity Plan 203
Implementing Information Security Policies 203
Information Access Policies 203
Protecting Private Data 207
Information Security Best Practices 208
Privacy Best Practices 208
Authentication and Authorization 208
Auditing and Logging 209
File Management 209
Error Handling 209
Secure Software Development Life Cycle 209
Secure Incident Management 210
Database Level Security 210
Sharing the Data with External Systems 210
Security Awareness and Training 210
Security Testing 211
Cloud Testing 211
Chapter Summary 212

Part IV: Infrastructure and NFR for the Banking Experience Platform 213

Chapter 8: Quality Attributes and Sizing of the DXP 215
Key Quality Attributes of DXP 215
Quality Attributes Deep Dive 217
Usability Requirements 217
Security Requirements 218

TABLE OF CONTENTS

    Reliability Requirements 219

    Scalability Requirements 219

    Availability Requirements 220

    Archival and Retention Requirements 221

    Logging and Auditing Requirements 221

    Performance Requirements 222

Infrastructure Sizing of DXP 222

Cloud Hosting of DXP Solution 224

    Tiered Architecture 224

    Cloud Deployment Considerations 225

    Cloud Deployment Model 226

Disaster Recovery and Business Continuity for DXP Applications 228

    DR Planning 229

    DR Implementation 230

    DR Maintenance 231

    DR Strategy Document 232

Chapter Summary 233

Chapter 9: DXP Performance Optimization 235

DXP Performance Optimization of Presentation Layer 235

    User Experience 235

Performance Testing for DXP 238

    Performance Testing Activities 238

    Key Performance Metrics 243

Performance Testing Framework 244

    Identify Critical Transactions 245

    Document Workload Model 245

    Qualitative Assessment 245

    Quantitative Assessment 246

    Predict 247

Performance Debugging Framework 247

    Identify the Root Cause 247

Optimize the Component/System/Layer	251
Common Performance Problem Pattern	252
Performance Case study	254
Application Context and Background	254
Performance Analysis	254
Recommendations and Improvements	256
Chapter Summary	258

Chapter 10: Transforming Legacy Banking Applications to Banking Experience Platforms ... 261

- Key Tenets of a Banking Experience Platform 262
  - Attributes of a Next-Generation Digital Bank 263
  - DXP Features for Next-Generation Digital Bank 265
- Main Trends in Digital Banking 268
  - Technology-Related Trends 268
  - Business Process-Related Trends 269
- Digital Transformation of Traditional Banks to Digital Banks 269
  - Reference Technology Architecture for a Digital Bank 269
  - Reference Functional View of Digital Bank 273
  - Main Digital Transformation Methods 278
  - Digital Transformation Road Map 288
  - Reimagining the Digital Banking Experience 288
- Chapter Summary 294

Part V: End to End Case Study ... 297

Chapter 11: End to End DXP Case Study ... 299

- Drivers and Key Requirements of the Dealer Platform Case Study 299
- Architecting the Next-Generation Dealer platform 300
  - Pain Point Analysis in Current Systems and Processes 300
  - Solution Tenets of Next-Generation Dealer Platform 302
  - Solution Design Principles 304
  - Persona-Based Information Architecture 307

    Persona-Based Design and Information Architecture 308

    Functional View of the Next-Generation Dealer Platform 310

    Seamless and Optimized Business Process 312

    Open-Source-Based Next-Generation Deal Digital Platform 313

Innovations and Next-Generation Technologies in Dealer Platform 318

Chapter Summary 320

Appendix A: Open-Source Tools and Frameworks 321

    HTTP Accelerator 321

    Web Server 321

    CSS Framework 322

    Scripting Framework 322

    User Interface Management 323

    Integration 324

    Application Server 324

    Server-Level Cache 325

    Content Management Systems 325

    CMIS 326

    SQL Database 326

    NoSQL Database 326

    IoT Framework 327

    Distributed Data Streaming 327

    Analytics Engine 327

    Distributed Processing 328

    Machine Learning Library and Framework 328

    Blockchain Frameworks 329

    Augmented and Virtual Reality 329

    Enterprise Search Engine 330

    Containerization 330

    Containerization Orchestration 331

Source Code Management	331
Continuous Integration and Continuous Delivery	331

Appendix B: Sample Code 333

User Interface	334
Integration	335
Data Mocking	336
Implementation and Logic	336
Deployment	337
Development	337
Production	338
Prerequisite	338
API Specification and API Mocking	339
Swagger-UI	339
Swagger-Editor	340
Swagger-Server	342
UI Screen Mocking on Node-RED	342
Apache Camel	346
Build Automation System	347
Run the Integration Application	354
Angular	355
Microservices Architecture	357
Microservices Components	358
Docker	363
Components	363
Summary	364

Appendix C: Further Reading 365

Index 367

About the Authors

Shailesh Kumar Shivakumar is an author, inventor and working as Practice Lead & Senior Technology Architect at Digital Practice of Infosys Limited. He is an award-winning digital technology practitioner with skills in technology and practice management and has experience in a wide spectrum of digital technologies, including enterprise portals, content systems, enterprise search, and other digital technologies. He has more than 17 years of industry experience and was the chief architect in building a digital platform that won the "Best Web Support Site 2013" global award. His areas of expertise include digital technologies, software engineering, performance engineering, and digital program management. He is a Guinness world record holder of participation for successfully developing a mobile application in a coding marathon.

Shailesh is deeply focused on enterprise architecture, building alliance partnerships with product vendors, and has a proven track record of executing complex, large-scale digital programs. He successfully architected and led many engagements for Fortune 500 clients of Infosys and built globally deployed enterprise applications. He also headed a center-of-excellence for digital practice and developed several digital solutions and intellectual property to accelerate digital solution development. He led multiple thought-leadership and productivity improvement initiatives and was part of special interest groups related to emerging web technologies at his organization.

Shailesh was awarded the prestigious "Albert Nelson Marquis Lifetime Achievement Award 2018" for excellence in technology and has received numerous honors and awards. He has won multiple awards including the prestigious Infosys Awards for Excellence 2013-14 "Multi-talented thought leader" under the "Innovation – Thought leadership" category, "Brand ambassador award" for MFG unit, "Best employee award", delivery excellency award and multiple spot awards, and received honors from executive vice chairman of his organization. He is featured as an "Infy star" in the Infosys Hall of fame and recently led a delivery team that won the "best project team" award at his organization.

ABOUT THE AUTHORS

Shailesh holds a Bachelor in Engineering in Computer Science and Engineering and is currently pursuing a doctoral degree in Computer Science. Shailesh has completed an executive management program from the Indian Institute of Management, Calcutta. Shailesh holds numerous professional certifications such as TOGAF 9 certification, Oracle Certified Master (OCM) in Java EE5 Enterprise Architect certification, IBM Certified SOA Solution Designer, and IBM Certified Solution Architect Cloud Computing Infrastructure. He is the sole author of four technical books on digital technologies, which were published by reputed publishers, and has published twelve technical white papers related to digital technologies. Shailesh is the sole inventor of two granted US patents (US9613341 and US10108601) and holds two more patent applications, and is a frequent speaker at events such as IEEE conferences and Oracle JavaOne. Shailesh has also published more than 10 research papers in various international journals.

Sourabhh Sethii is a Technology Analyst at Infosys Technologies Limited. His areas of expertise include Blockchain, Internet of things (IoT), machine learning (ML), Java enterprise technology, front-end frameworks, and integration technology. He has hands-on experiences with many technologies like database integration, continuous integration, and security, along with performance analysis and web frameworks like Angular and Node. Sourabhh has worked on multiple domains such as banking, finance, and manufacturing. He has achieved multiple honors like "Most Valuable Player," "Insta Awards," and "Best Employee Award" from the heads of his unit in Infosys. He has published many technical white papers.

Sourabhh holds Master degree in Software Systems specialized in Data Analytics from Bits Pilani, Rajasthan, India.

About the Technical Reviewers

George Koelsch is a retired system engineer who resides in West Virginia, after 33 years in the DC metro area. He started system engineering 42 years ago while in the US Army, and had continued that work for an additional 33 years as a contractor for the Federal Government. With a five-year stint as an Industrial Engineer at Michelin Tire Corporation, he learned to become an efficiency expert, which he then applied to system engineering and project management to tailor the lifecycle development process before his contemporaries in the DC area were doing it. In his spare time, he authored ten nonfiction articles on computers, coin collecting, stamp collecting, and high-energy physics. Apress published his book titled *Requirements Writing for System Engineering* in October 2016. He now focusses on writing, all his hobbies, and other projects he has time to work on now.

Venkata Kakarlapudi is a Senior Technology Architect at Infosys Limited, with over 15 years of industry experience. His areas of expertise include Java, enterprise portals, and Web content management systems. He has experience of implementing multiple large-scale enterprise applications for Fortune 500 companies across geographies. He previously headed a center of excellence for enterprise portals at the digital experience practice in his organization and is part of the core architecture team. He holds an engineering degree in Mechanical Engineering and has completed an executive management program for IT executives from the Indian Institute of Management, Bangalore.

Acknowledgments

Shailesh would like to acknowledge and thank Verma V.S.S.R.K for their valuable inputs and review comments. Shailesh would also like to recognize and thank Dr. P. V. Suresh for his constant encouragement and immense support.

Sourabhh would like to thank his parents (Ritu Sethi and Sat Dev Sethi) and brother (Shrey Sethi), who were the guiding light behind him.

The authors also like to immensely thank Mr. George Koelsch for his technical review; his feedback has added great value to the book.

The authors want to sincerely acknowledge and thank profusely the Infosys team that includes the managers Jitin Singla, Saumitra Bhatnagar, Vivek Rastogi, Sarweshwar Panda, Sumit Arora, Aditya Kumar Soni, our colleagues and our friends who have facilitated us in accomplishing this task. Special thanks to Rahul Krishan for precious guidance and support. The authors would also like to convey sincere thanks to our mentor and friend Jasleen Khokhar, who read the manuscript at different stages as it evolved from shoot to bud, from flower to fruit. The authors would like to extend thanks to Anchit Madaan (Blockchain Core Team), Deepak Garg, Himanshu Arora, Jaskirat Singh, Babu Krishna Murthy, Kiran Korke, Nishant Satija (Digital Experience Team), and Arpit Kulshrestha.

Our special thanks to Shivangi Ramachandran, Rita Fernando, and the editors, technical team, designers, and publishing team at Apress for providing all necessary and timely support in terms of review, guidance, and regular follow-ups.

Introduction

As enterprises embark on their digital transformation journey, they define the vision, road map, and objectives of the digital transformation programs. Digital transformation involves legacy modernization, reimagining digital experiences, implementing cloud-first and mobile-first models. Such digital transformation involves various challenges and risk factors including but not limited to niche technology stack, unavailability of skilled resources, long time to market, and such. Enterprises need to carefully evaluate technology trends and future outlook, and invest in the technology stack that caters to their current digital transformation goals as well as their long-term digital vision.

Digital experience platforms (DXPs) are an integrated set of technologies and tools that provide best-of-breed modern digital technologies for enterprises. A DXP has a preintegrated set of technology stacks that addresses the risk related to a niche/unproven technology stack and risky integrations. DXPs are designed on the platform philosophy so that they can be easily extended and scaled to meet future demands of scalability and onboard new innovations. A DXP is one of the most popular approaches for building an enterprise grade digital platform. A DXP provides a set of capabilities to quickly develop a personalized, secure, and scalable enterprise platform. DXPs are designed in such a way that they can incorporate modern digital technology to build next-generation enterprise applications.

You can develop your own digital experience platform. The book looks at various open-source tools, technology, and frameworks that can be used for building DXPs. This book covers core concepts to build enterprise grade DXPs. Readers get a holistic view to build DXPs and will be able to transform existing applications to a DXP that is capable of incorporating emerging technologies in near future.

DXPs are not just limited to a few commercial products. Enterprises can build their own experience platforms to meet their needs. This book discusses the methods and technology frameworks across various layers to design a DXP.

We divided this book into five parts: requirements and design, development, security, infrastructure/NFR, and a case study to cover end-to-end DXP lifecycle scenarios. We discuss proven best practices, design methods, and technology frameworks along with contextual real-world case studies for each of the chapters.

INTRODUCTION

In the requirements and design part, we introduce various concepts of DXPs and elaborate on requirements elaboration methods. We also provide an in-depth discussion of various design elements of DXPs such as UI design, integration design, and such. The chapters in this part cover the requirement gathering phases, functional requirements, and sample use case to develop your own DXP application; user experience requirements to develop your own user interface and mobility requirements to develop your own mobile experience; nonfunctional, social and collaboration, security, infrastructure, disaster recovery, and rollout requirements to develop your own digital experiences platform. This is the first step to develop and analyze the requirements to build an enterprise DXP. The design chapter covers the patterns and architectural strategy along with various layers of the DXP. This chapter also briefly discusses the integration of various emerging technologies such as IoT integration design, Blockchain design, big data, and NoSQL design, and AI automation design along with chatbot case studies, enterprise search engine capabilities, and introduction of augmented reality with DXP applications, along with recent trends in CICD (continuous integration and continuous deployment) using application containerization technique.

The development part mainly discusses the detailed design and development of DXP layers such as the user interface layer and integration layer. The chapters in this part cover each and every aspect of developing the user interface using open-source web frameworks, modular UI components and key features, integration of UI components using open source ESB and integration frameworks, UI development lifecycle and best practices, along with a BXP (Banking experience platform) case study.

In the security part, we provide an elaborate discussion of information security and overall security of DXPs. The chapters in this part cover the concepts and best practices while developing an application's security, along with information security policy and principles.

The infrastructure/NFR part discusses various quality and nonfunctional attributes such as performance, infrastructure sizing, and such. The chapters in this part cover the NFR(nonfunctional requirements), that is, scalability, availability, performance, modularity, extensibility, and security of the DXP's application along with quality attributes such as usability, configurability, stability, interoperability, efficiency, flexibility, and maintainability of the platform.

Finally, we wrap up with an elaborate digital transformation case study of a legacy system in the last chapter. The case study chapter provides insights into the digital transformation of a legacy application to a Digital experience platform. It covers concepts

INTRODUCTION

like gamification, predictive analysis, dashboards, and chatbots; and technologies like artificial intelligence, Blockchain, and augmented reality are discussed in brief.

The book can be used as a reference while using any existing DXP tools or for developing a new DXP from the ground up. Digital practitioners, web developers, and digital architects can leverage the best practices, methods, and technology frameworks discussed in this book.

PART I

Requirements and Design

CHAPTER 1

Introduction to Digital Experience Platforms

The digital strategy of all organizations primarily focuses on providing rich and engaging user experience. Customer experience-focused strategy leads to increased customer engagement, which in turn increases key success metrics such as site traffic, repeated visits, conversion rate, and such.

Digital experience platforms (DXPs) provide an integrated set of technologies built on platform philosophy to engage users throughout their journey. DXPs provide seamless user experience across all user touch points. A DXP is a convergence of all customer-centric technologies such as content management systems, portals, analytics, campaigns, targeting, search, mobile apps, and such.

Industries dependent on digital technologies are undergoing rapid disruption mainly fuelled by changing tech-savvy customer expectations, disruption in digital technologies, and due to widespread popularity of mobile devices. Incumbent organizations are undertaking digital transformation exercises to meet the customer expectations and to stay competitive.

Organizations can increase their online revenue through user engagement. User engagement also increases cross-sell and upsell opportunities, and increases user retention and lifetime value of a customer.

CHAPTER 1   INTRODUCTION TO DIGITAL EXPERIENCE PLATFORMS

Boundaryless Banking Enabled by Digital Technologies

Tech-savvy banking customers expect the banking experience to match or surpass the best experience of social media platforms, hence it is imperative for banks to understand the trends and enhance the banking experience. Digital banks enable a boundaryless and physical branchless experience supporting these features:

- Mobile-first strategy enabled through mobile apps or mobile web platforms
- Omnichannel experience (a seamless user experience on all supported devices and browsers) to provide optimal user experience on all access devices
- Seamless and simplified processes across all touch points throughout the user journey
- Relationship oriented by rewarding loyalty and sustaining long-term partnership with customers
- Responsive to market disruptions, changing customer demands, and other requests
- Digitized business models to foster the innovation
- Rapid innovation in adding digital capabilities

In the subsequent section we will briefly discuss DXPs.

Overview of DXP

DXPs are primarily user-centric engagement platforms that provide a unified view, with rich user interface for enhanced end-user experience. DXPs provide a platform-based approach to enable all the needed digital capabilities. In this book we explore various aspects of a digital experience platform such as user experience design, integration, security, and such. In this regard we will explore the concepts of DXP in understanding the background for using a DXP to build a banking experience platform.

In this section we will provide details of the DXP.

Key Tenets of a DXP

The key tenets of DXPs are defined as follows:

- Platform orientation with an integrated set of technologies that provides capabilities for presentation, content management, commerce, marketing search, analytics, campaigns, and such. The platform model should also support future extensibility.

- Lean and agile platforms with lightweight integration components. A lean model includes lightweight user experience integrated with lightweight service components.

- An integrated and personalized view to provide a holistic view of all customer activities across all touch points. This can be achieved by information aggregation from multiple information sources and delivering personalized experiences.

- Provide software as a service (SaaS) and cloud deployment option to provide the digital experience as a service.

- Provide an integrated experience catering to various business channels such as marketing, sales, and services.

- Self-service for end users and for business stakeholders to improve user experience and productivity.

- Agility in developing new features and implementing changes for responding to changing market demands.

DXP Reference Architecture

The reference architecture provides the core services and components that are used in a typical digital experience platform. The services and components enable the needed business capability for the application using the DXP; we will elaborate each of these components in detail shortly.

DXP reference architecture is shown in Figure 1-1.

CHAPTER 1 INTRODUCTION TO DIGITAL EXPERIENCE PLATFORMS

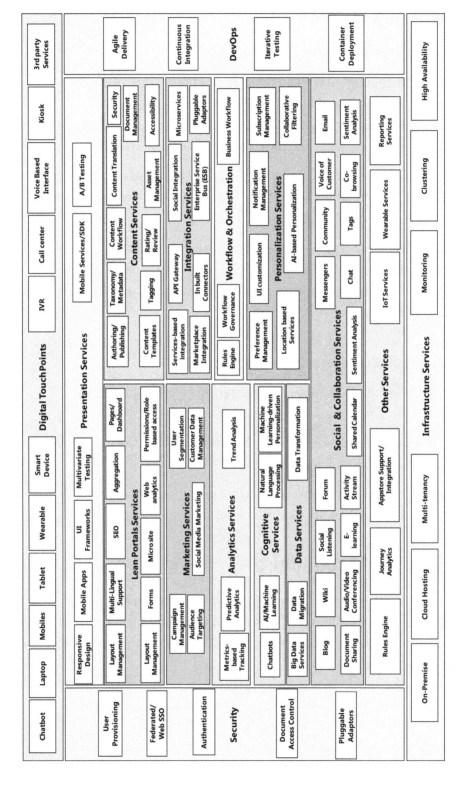

Figure 1-1. DXP reference architecture

The core components of a typical DXP platform are as follows. We have identified the core components in each of the layers:

- *User touch points*: This layer consists of various digital touch points the end user uses during the journey. The end user could use smartphones, desktops, tablets, third party services, or wearable and such devices to access the DXP services. Users expect device-optimized, seamless and personalized information access across all digital channels. All user access channels and devices come in this layer.

- *Presentation services*: The DXP provides various presentation services to cater to a wide variety of digital touch points. This includes mobile apps for smartphones, UI frameworks, and responsive design for mobile web applications, web services for third party consumer, and A/B testing for presentation testing. Presentation services are mainly responsible for defining the user interface and user experience. We elaborate presentation services and user touch points in Chapter 4.

- *Lean portal services*: In this category, the portal provides various complementary presentation capabilities such as personalized experience (user experience based on end user preferences and past history), consistent branding, unified view, forms (for user registration, queries, and such), search engine optimization (SEO) (to make web pages more visible), multilingual presentation, and such.

  - Lean portal services provide business-friendly controls to manage pages (page creation, layout, web analytics, navigation) and brands.

  - Lean portal services provide a single-stop-shop view of personalized content by aggregating information from various sources.

  - Web analytics provide vital real-time customer insights, and help in understanding customer activities and interests. These insights can be used for customer segmentation, trend analysis, and targeted content delivery/contextual recommendations.

- *Content services*: In this category, the DXP provides various content management services such as content authoring, content tagging, content publishing, content translation, and such. As the DXP provides an integrated set of features, support for various content types, content administration, content templates, content metadata, and other content related services will also be provided by the DXP. Other complimentary functionality such as document management services, digital asset management (DAM) services, content workflows, and metadata management are also included in this category.

  - Content services provide content lifecycle management features (content creation, content updates, content publishing, content translation) and support a wide variety of content.

  - Content services provide other features such as rich text editor, content workflows, and such.

- *Campaign and marketing services*: One of the core features of a DXP is to enable digital marketing campaigns. To provide this, the DXP includes features such as campaign management (defining, launching, and monitoring campaigns), audience targeting (sending targeted information to the relevant audience), social media marketing, user segmentation (grouping users based on their interests, access patterns, and such), and customer data management (unified management of customer data across all customer touch points).

  A DXP provides campaign management features (campaign creation, campaign targeting) and user segmentation (categorizing end users based on demographics, interests and such) in this category. Customer data management (profile data, preferences data, transaction data, and navigation data) is used for understanding customers and provides targeted content. Customer data is used to provide a single view of customer data (activities, preferences, transactions, feed, etc.) in the dashboard. Other marketing functionality such as social media marketing is included as well.

  - Campaign and marketing services mainly deliver targeted content based on user attributes, preferences, analytics, and such.

- *Analytics services*: This includes web analytics-based tracking using predefined metrics, trend analysis, and predictive analytics.

- *Integration services*: Enterprise integration is the most significant component of a DXP. In order to aggregate information from various information sources to provide a unified view, a DXP should support a variety of integration formats and should provide flexible and extensible integration features. Hence a DXP offers standards-based integration methods such as API support, modular services, services support, and plugin support. Most of the DXPs offer built-in support for microservices, REST (Representational State Transfer) and JSON (JavaScript Object Notation)-based services and adaptors for most popular enterprise interfaces (such as databases, enterprise resource planning [ERP], etc.)

    - The in-built adaptors and integrators improve the productivity of end users and optimize the return on investment (ROI) of the DXP.

    - DXPs provide standards-based integration options (such as REST-based integration, web services, and such), which can be leveraged for integrating with new products and technologies.

- *Social and collaboration services*: In this category, a DXP provides various collaboration features such as forums, blog, wiki, chat, knowledge base, messengers, communities, calendars, email, and such. These features enable end users to share the information and facilitate a self-service model. The social capabilities enable users to collaborate, harness collective intelligence, socialize, and improve productivity.

    - Social and collaboration enable users to collaborate and engage customers at social touch points.

- *Workflow and orchestration*: DXPs enable designing and implementing agile, automated, and dynamic business processes through workflow modeling, a configurable rules engine, and workflow governance.

- *Search services*: Information discovery is mainly enabled through search features such as site search, content search, and federated search. DXPs also support advanced search features such as result filtering or faceted searching.

  - Search services improve the user productivity through efficient information discovery.

- *Commerce services* (Optional): Based on the usage domain, DXPs also provide various commerce services such as catalog management, order management, product information management (PIM), inventory management, etc.

- *Cognitive services* (Optional): In this category we have services that leverage artificial intelligence (AI) and machine learning and natural language processing methods to provide personalization recommendations based on insights gathered.

- *Data services* (Optional): This includes services related to data processing such as Big Data services, data migration-related services, and data transformation-related services.

- *Infrastructure services*: In this category, a DXP offers various features such as support for on-premise deployment, cloud deployment, container deployment (deploy code base to run independently for increased robustness and failover), and multitenancy (a single codebase used for multiple-user groups). A DXP also supports other high availability features such as clustering, monitoring, etc.

- *Workflow and orchestration services*: These services are mainly used for orchestration of business processes. This category includes components such as rules engine, workflow governance, and business process modeling tools.

- *Personalization services*: Personalized delivery is an essential feature of any DXP. This module includes preference management, UI customization (ability for user to customize widgets, page layout), notification management (alerting and notifying users), subscription management (enabling and disabling of subscriptions for the user), collaborative filtering (recommending products based on their attributes, behavior of similar customers, and such), and AI-based personalization (personalized based on matching learning of users' interests and activities).

- *Security*: In this category, a DXP offers various authentication and authorization features such as support for an access control list, public–private key infrastructure, web SSO (single sign-on); pluggable adapters, Lightweight Directory Access Protocol (LDAP) integration, Security Assertion Markup Language (SAML) integration, and NT LAN Manager (NTLM) integration. We discuss elaborate security features in Chapters 6 and 7.

- *DevOps*: DXP methodology also supports and uses various open-source DevOps features such as Agile Delivery (such as Agile Management with Slack or Jira); continuous integration or CI (such as Jenkins); iterative testing; container deployment (Docker or Kubernetes); etc.

- *Other Services*: It can also support third-party integration of open-source features available in the market, such as Rules Engine, Journey Analytics (Google web Analytics, Open Web Analytics, similar web, etc.), Appstore support/integration (Google Playstore/Apple appstore), IoT services (Iotivity), wearable services, and reporting services.

Evolution and Drivers for DXP

In this section we discuss the various stages of evolution of digital platforms and the key drivers of DXPs.

Evolution of Digital Platforms

Various stages of evolution of digital platforms are shown in Figure 1-2.

CHAPTER 1   INTRODUCTION TO DIGITAL EXPERIENCE PLATFORMS

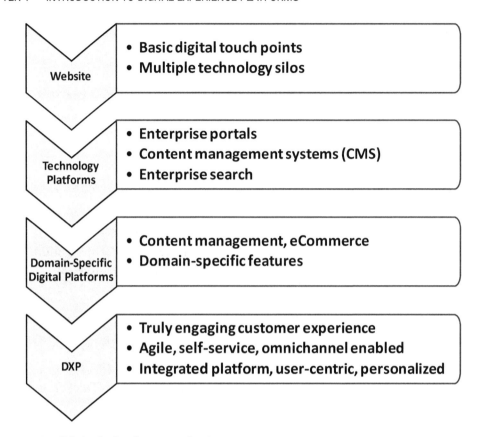

Figure 1-2. Digital platform evolution

The evolution of the digital eco system is depicted in Figure 1-2. During initial stages, web sites were mainly used for information delivery. Web sites needed to be integrated with multiple backend systems and services needed for the business. The next stage of the digital ecosystem was technology platforms such as enterprise portals, CMS, search engines, analytics engines, and such. These technology platforms addressed specific concerns of the enterprise applications. For instance, enterprise portals mainly addressed concerns related to presentation, information aggregation, and personalization; CMS managed the end-to-end lifecycle of content and search engines handled the indexing and searching related concerns. In this scenario we needed multiple enterprise products to build a digital solution. The next step in the evolution was domain-specific digital platforms. For instance, CMS-specific digital platforms provided basic presentation, basic search and ready-to-use integrators/plugins for search engines, and campaign management systems. Similarly, e-commerce platforms provided storefront portals and basic content management capabilities. These prebuilt/out-of-the-box capabilities built around the core capabilities reduced the number of products and technologies that need to be integrated.

CHAPTER 1 INTRODUCTION TO DIGITAL EXPERIENCE PLATFORMS

Digital experience platforms are the next step in the evolution journey. DXPs provide a preintegrated stack to use and extend for any enterprise digital solution.

Some of the challenges addressed by DXPs are shown in Table 1-1.

Table 1-1. *DXP Challenges and Solutions*

Challenge	Description	How DXP Addresses It
Technology complexity for building enterprise application	• Too many products and technologies adding to the overall technology complexity. • Too many integrations involved products.	• DXPs provide preintegrated stack and provide all necessary capabilities for building enterprise digital solutions.
Performance and availability challenges	• Traditional web platforms tend to be "heavy" for installation and maintenance, and pose performance challenges. • Too much integration is also causing performance, scalability and availability issues.	• DXPs are built on a lean model offering Lean UI frameworks/products for rich user experience needs. • Cloud-native DXPs are well equipped to handle the availability and on-demand scalability challenges.
Productivity challenges	• The inherent complexity of the traditional platforms is resulting in greater time for implementation. • Missing common, reusable component and frameworks.	• DXPs can be configured and extended to build needed capabilities. • Preintegrated technology stack and development tools improve productivity.
Not aligned with overall digital strategy	• Traditional digital technologies and products pose challenges in fully implementing digital strategy elements such as mobility enablement, analytics, and cloud	• DXPs can be used to implement digital strategy, as it provides all the necessary building blocks of digital technologies with extensible architecture. • DXPs support and provide modern digital technologies such as collaboration tools, AI tools, and mobile apps.
High maintenance and infrastructure cost	• High software licensing and support costs. • High development, testing, and skill costs involved.	• A DXP is a single, modern product offering combined functionalities in an overall digital space.

13

CHAPTER 1 INTRODUCTION TO DIGITAL EXPERIENCE PLATFORMS

Business Drivers for DXP

Listed in Table 1-2 are the key business drivers across various industry verticals. We will elaborate on these in the contextual case studies in coming chapters.

Table 1-2. *DXP Drivers and Business Scenarios*

Vertical/Industry Domain	Drivers and Key Business Scenarios
Retail and digital commerce	• Lightweight and agile platforms with hyperpersonalization, catering to customer intent to drive conversions • User journey optimization and user engagement across all digital touch points and access channels • Integration of online (chat, call center, web, mobile, and offline channels (brick-and-mortar stores) • AI (artificial intelligence) and machine learning-based predictive analytics, recommendations, and personalization • Unified view of customer actions and information • Other features are conversational commerce, virtual assistants, augmented reality, marketplace integration, API-based backend integration (also known as "headless" mode), and multitenancy model.
Marketing and sales	• Campaign management • Personalized/targeted content delivery • Customer segmentation • Configurable and extensible product • Social marketing • Brand management • Micro site management • Campaign management

(*continued*)

Table 1-2. (*continued*)

Vertical/Industry Domain	Drivers and Key Business Scenarios
Insurance	- Product comparison tools - Mobile apps that provide services related to quotes, claims, service requests, policy management - Customer self-service related to policy management, service requests, payment and profile management - Advanced analytics for fraud detection - Campaign management, lead management
Manufacturing	- Convergence of B2C (business to consumer) and B2B (business to business) channels - Information delivery - Collaboration - Unified dashboard view
Customer Service	- Self-service - Knowledge management - Search and information discovery - Collaboration tools and collective intelligence harnessing tools such as wikis, blogs, forums, communities - AI-based virtual personal assistants (VPAs), chatbots - Machine-learning based contextual recommendations - Social media integration
Common business drivers across all verticals	- Lower operational cost and continuous improvement - Data-driven decision making - Self-service - Faster time to market through agile delivery and maximal reuse of out-of-the-box platform components - Agile delivery - Frictionless and automated processes and improved productivity

We have detailed various digital transformation case studies and scenarios for a DXP. In subsequent sections we will look at the details of banking experience platform as an example implementation of a DXP. DXP principles can be applied to design other experience platforms as well.

Overview of Banking Experience Platform

We will look at the core features of the banking experience platform.

> **Note** The banking experience platform referred to in this book mainly refers to the retail/consumer banking solution used by banking customers.

Key Tenets of Banking Experience Platform

The primary motivation of a banking experience platform is to provide a holistic and engaging user experience for the online customers of the bank. This book elaborates on building a banking experience platform to fulfill this vision. The primary tenets of a banking experience platform are as shown in Table 1-3. We will elaborate on these platform tenets in Chapters 3, 4, and 5.

Table 1-3. Key Tenets of a banking Experience Platform

Key Tenets of banking experience platform	Attributes
Integrated view	Single-stop-shop experience, aggregation of data from all interfaces
Personalized experience	Relevant and contextual information delivery, role-based access
Intuitive user experience	Interactive and response user interface, omnichannel access
Self-service	Enhanced information discovery, smart search, comparison tools, calculators, decision-aiding tools
Secured transactions	Easy to use, and simplified and secured transactions

CHAPTER 1 INTRODUCTION TO DIGITAL EXPERIENCE PLATFORMS

The main features of a future-state banking XP is depicted in Figure 1-3.

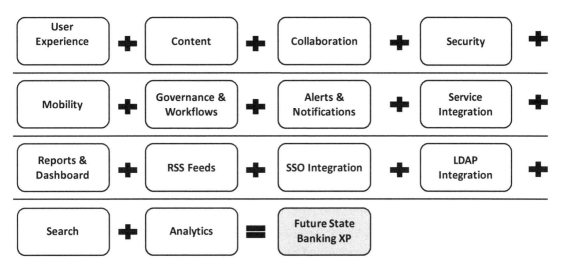

Figure 1-3. *Key features of a modern banking experience platform*

As depicted in the diagram, a future state banking XP needs a responsive and omnichannel-enabled use experience, engaging content. Collaboration features such as chat, blogs, and forums enable active participation of end users. The banking XP should ease the integration with security systems and provide service-based interfaces for consumers. The platform should provide timely alerts and intuitive visualizations (charts, reports, dashboards) to help the customer in decision making. Search is the key information discovery tool for the banking XP. Analytics enablement helps in personalization of the user experience.

High-Level Requirements of Banking Experience Platform

Table 1-4 shows the high-level requirements of a banking XP categorized into main categories. We discuss these requirements in Chapter 2.

CHAPTER 1 INTRODUCTION TO DIGITAL EXPERIENCE PLATFORMS

Table 1-4. *High-Level Requirements of a Banking Experience Platform*

Requirement Category for Banking XP	High-Level Requirements
User experience	Customer-centric design, lightweight/lean UI, responsive UI elements, mobile apps, dashboard UI, simple and easy-to-use interfaces, high usability, intuitive information architecture, personalization
Security	Authentication, authorization, SSO, flexible login methods, multifactor authentication, adaptive authentication, auditing
Enterprise integration	Integration with needed enterprise interfaces such as core banking systems, commerce platforms, ERP systems, enterprise services and interfaces, light-weight services
User engagement features	Personalization, profile management, gamification (using gaming concepts such as points, scores, and badges to enhance and encourage user participation), contextual content delivery, social media integration, self-service tools, analytics, collaboration (chat, blog, wiki, forums, document sharing, etc.), dashboard views, profile management, review and rating, alerts/notification, localization and such
Self-service tools	Reports, search, calculator (related to loans, policies, etc.), tools (such as risk profiling tool, budget planning tool, need analysis tool), incident management
Optimized processes	Quicker registration, account opening process, bill payment process, funds transfer process
Analytics	Customer behavior insights, navigation patterns, metrics-based tracking, user segmentation,
Anytime anywhere availability	Cloud enabled, platform accessible on all devices and browsers, disaster recovery capabilities
Content management	Campaigns, web content, promotion content, publishing, content workflows

CHAPTER 1 INTRODUCTION TO DIGITAL EXPERIENCE PLATFORMS

The sample objectives for a banking XP are shown in Figure 1-4.

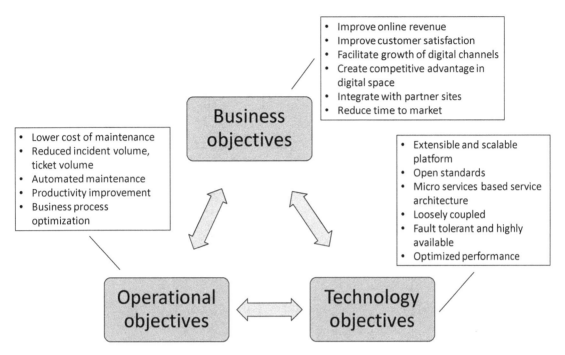

Figure 1-4. *Key objectives of a banking experience platform*

Sample reference architecture of a banking XP is given in Figure 1-5.

CHAPTER 1 INTRODUCTION TO DIGITAL EXPERIENCE PLATFORMS

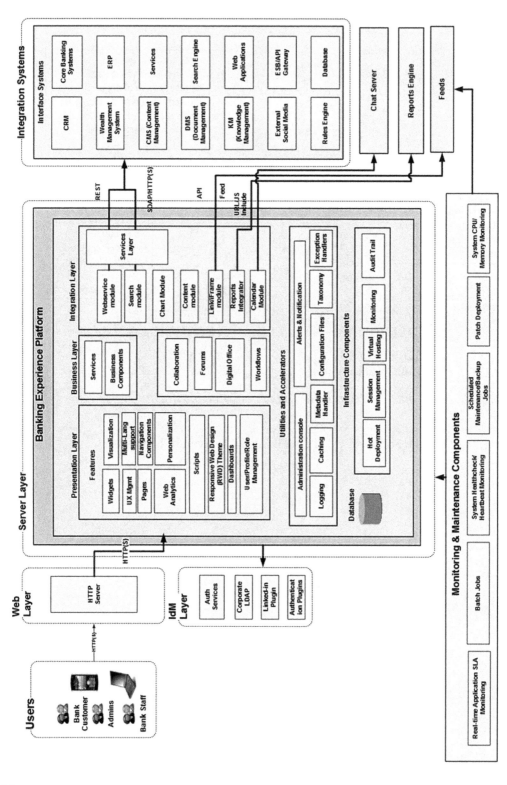

Figure 1-5. Reference architecture of a banking experience platform

The banking XP is typically a layered system providing a loosely coupled platform. The platform will be used by customers, admin, and bank staff.

The core platform consists of three layers: presentation layer, business layer, and integration layer.

The presentation layer provides user interface components such as widgets, pages, web analytics, localization, personalization, and visualization components. These components define and impact the end-user experience.

The business layer consists of core business components to implement the business logic, transactions, and workflows and provide collaboration features.

The integration layer provides integrators and plugins to interact with enterprise interfaces through lightweight services. Enterprise interfaces include customer relationship management, wealth management systems, content management system, document management system, search engine, enterprise services, enterprise resource planning, core banking system, enterprise service bus, etc.

The platform also provides various utilities and accelerators such as loggers, caching components, and taxonomy and exception handlers.

Three Ps of BXP

The three Ps of BXP are as follows:

- *Purpose*: Customer satisfaction, customer retention, growth of digital services, integration with other partners, reduce time to market.

- *Process*: Banking organization assessment of each major value stream to make sure each step is valuable, capable, available, adequate, flexible, and that all the steps are linked by flow, pull, and leveling.

- People: Bank customers, admin, and bank staff.

Sample Technical Capabilities of Banking Experience Platform

Table 1-5 is a sample list of technical capabilities for a banking XP:

CHAPTER 1　INTRODUCTION TO DIGITAL EXPERIENCE PLATFORMS

Table 1-5. *Technical Capabilities of a Banking Experience Platform*

Capability Group	Capabilities
Presentation	- Mobile-first components - Responsive and interactive user interaction elements - User experience management - Consistent user experience across channels - Intuitive dashboard providing unified view of transactions across all channels - Intuitive visualizations (charts, graphs, reports) - Omnichannel access - Web analytics - A/B testing - Web analytics
Personalization	- User profiling - Segmentation - Analytics-driven personalization - Targeted marketing
Security	- Identity management - Single sign-on (SSO) - Encryption - Adaptive security - Multifactor authentication - Fraud detection - Flexible authentication methods - Data security

(*continued*)

CHAPTER 1 INTRODUCTION TO DIGITAL EXPERIENCE PLATFORMS

Table 1-5. (*continued*)

Capability Group	Capabilities
Web content management	- Common content types, templates, and workflows - Taxonomy and metadata management - Content preview - Content Versioning and life cycle management - Federation and syndication - Digital marketing
Intelligent information access	- Indexing - Search - Information access administration - Federated search - Standard connectors - Content query
Integration	- Managed service consumption - Information aggregation from all relevant data sources (transactions, risk score, etc.) - Data consumption - Information aggregation across all channels and touch points - Transactions
Transaction management	- Secured transaction manager - Payment manager - Funds transfer

(*continued*)

Table 1-5. (*continued*)

Capability Group	Capabilities
Enterprise Integration	• Light weight services layer (microservices) • Services-based integration. • Partner integration (third-party API)
Social and collaboration	• Social listening • Forums, chats, communities • External social media collaboration • Training and self-learning content
Other features	• Transaction reporting • Search • Tools (forecasting tools, comparison tools, reporting tools, self-service tools) • Enhanced analytics (what-if scenarios analysis tools, smart recommendations, cross-sell insights) • Well defined governance and processes • Defining single source of truth • Incident management • Gamification • Cloud enablement • Use of AI features such as chatbot (rule-based chatbot or predictive-based chatbot), recommendation, data analytics, and such

Sample Key Performance Indicators of Banking Experience Platform

In order to measure the success of the banking XP, we use following key performance indicators (KPIs):

- *Increased business transactions*: The enhanced banking XP should make it easier for online banking customers to carry out transactions.

- *Improved user satisfaction*: The banking XP should improve the usability thereby improving the overall user experience.

- *Faster time to market*: The banking XP should provide optimized processes and reusable technical building blocks that reduce launch time of new releases.

- *Reliability, availability, and performance*: The banking XP should provide a reliable and highly available platform to provide an enhanced user experience.

Digital Imperatives for Modern Banks

Disruptions in digital technologies are impacting modern banks. Banks are heavily adopting digital technologies to remain competitive. Given below are the key digital imperatives for modern banks:

- A mobile device is the primary access channel for accessing web applications. So banks should provide mobile friendly web or apps for mobile platforms.

- Digital banks are one of the emerging trends, wherein the banks provide branchless digital banks that solely rely on digital channels.

- Increasing use of gamification concepts by rewarding banking customers for their online transactions and activities.

- Increasing use of AI tools such as chat bots, personalized recommendations, etc.

- Money management tools (such as spend analyzers, budget planner, debt analyzer) and self-service tools are increasingly used in digital banks to aid customer decision making and to minimize information clutter.

- Digital payments and digital wallets are some of the features gaining momentum in digital banking channels.

- Social banking that actively leverages social and collaboration features is one of the most popular features in the digital bank.

- Optimized and simplified processes (such as 1-step registration, or a simplified customer acquisition process)

CHAPTER 1   INTRODUCTION TO DIGITAL EXPERIENCE PLATFORMS

Summary

- Digital experience platforms (DXPs) manage the business processes and decrease time required from development to production.

- DXPs provide cost effective digital services to customers.

- Multiple technologies, lean structure, and digital touch points collaboratively enhance the overall digital experience. A DXP is a highly flexible way to develop as well as interact with digital services.

- The main idea or methodology behind the DXP is it has lean structure, which delivers maximum to the end customer with minimum cost on the resources.

- DXPs make the complete production and digital service delivery process efficient. Different technologies optimize and speed up the process.

- A DXP can manage employees and customers of organizations on a single platform.

- A DXP can be single point for organizations to digitize business process.

- A DXP provides holistic user experience to all the stakeholders of organization.

CHAPTER 2

Gathering Requirements

Requirement gathering is one of the critical functions to ensure the success of a system and platform. Digital Experience Platform (DXP) certifies the successful delivery of the project that covers omnichannel as well as crossed-channel requirements. A DXP focuses on developing and delivering a platform where an application is developed once and deployed everywhere. A DXP provides omnichannel capabilities by providing reusability of user interface (UI) components (also called widgets or portlets) and manageable content. A DXP supports native applications. Besides functionality typically delivered by these web-based UI components, content is made reusable between channels, that is, Web, mobile device, tablet, interactive voice response (IVR), and automatic teller machines (ATMs), etc. and platforms, that is, Web, Android, iOS, etc. as shown in Figure 2-1.

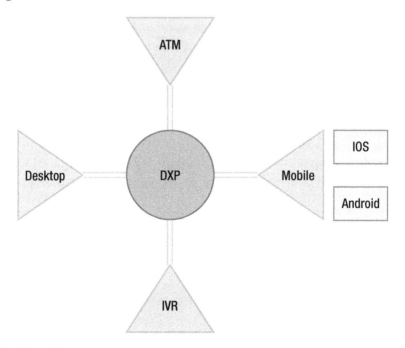

Figure 2-1. *Omnichannel*

CHAPTER 2   GATHERING REQUIREMENTS

The requirement elicitation and elaboration process enhances DXP application requirement gathering and analysis. Requirement elicitation and elaboration is the process of gathering and analyzing requirements. In this process, the developer along with the system engineer interact with the systems, stakeholders, documents, and case study of existing applications along with proof of concepts done by teams. These requirement elicitation and elaboration methods are best suited for gathering digital platform requirements, but there are other methods also that can be used for requirement elicitation and elaboration.

- *Requirement workshops*: Workshops are beneficial on the grounds that business analysts want to take stakeholders' opinions and consensus. Workshops could be combined with brainstorms for discovering requirements, where ground rules are predetermined at the outset of the workshop.

  - Workshops help business analysts to build mock-ups or prototypes for refining and validating requirements.

  - Workshops could also include a walk-through for reviewing requirements.

- *Stakeholder interviews*: Stakeholder interviews can lead toward success of the application, although stakeholders come in all shapes and sizes. The motive of the interview is to talk with those people who will spend most of the time using the things you plan to design, though it would help to first determine what that thing actually is.

- *Documentation study*: It helps to understand the existing systems and challenges, so that they are incorporated in the new system. For example, understanding existing systems and their configurations helps us to integrate the existing system's components with the

DXP application. It further assists in maintaining the integrity of the system, as well as aiding understanding of the requirements.

- *AS-IS system study*: You should go through other existing applications in the organization; it will further assist you to incorporate the existing service with the DXP applications. You should model the user journey and do pain point analysis, which help you to understand the pain points faced by customers in the existing system. Process study that is studying and analyzing the domain and process involved in the domain. Study activities include management, and the technical and professional people who are familiar with the processes involved in the systems. The gathered information is used to identify existing gaps, which helps you to optimize the business process.

- *PoCs (proof of concepts)*: Proof of concept is also known as proof of principle. A PoC is a small exercise to test the design or idea. PoCs help the team to understand functionality and estimate complexity, which will further aid in developing the DXP application.

- *Benchmarking competitor sites*: The purpose of benchmarking is to interpret and gain a level of insight that allows you to evolve a digital strategy based on competitor insight. It helps you to understand the potential customer and further aids in estimating performance and capacity-related requirements.

- *Questionnaires*: An effective questionnaire helps you to decide the actual user requirements. The answers can be used to analyze the results. When the number of stake holders is greater, or the resources and funds are less, then a questionnaire is the best method to analyze the requirements. The questionnaire should be well defined and effective. All the questions should be relevant to the requirements.

CHAPTER 2 GATHERING REQUIREMENTS

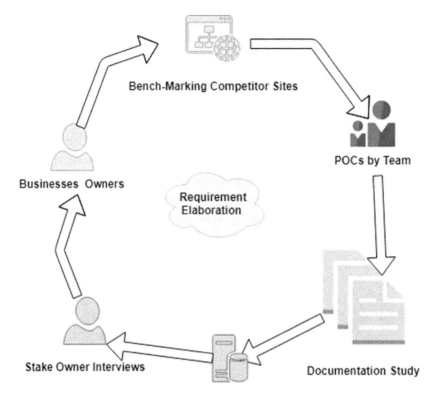

Figure 2-2. *Requirement elaboration*

Figure 2-3 will help you to understand the elaboration phases and associated requirements. For example, functional requirements can be gathered through requirement workshops, system study, and stakeholder interviews.

CHAPTER 2　GATHERING REQUIREMENTS

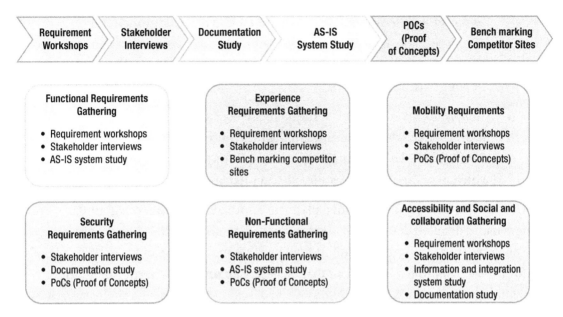

Figure 2-3. Requirement gathering

You will look into banking experience platform (BXP) use cases throughout this chapter. Now we would look into use case as well as user stories, describes the system interaction with the environment and actors; it contains a detailed description of the following objectives, which provide detailed descriptions about how a user interacts with the system and how the system will respond to the user's action.

- Actors
- Preconditions
- Postconditions
- Alternative paths
- Main scenario

A use case captures all the possible ways the user and system interact and a detailed description about goals and results. We prefer use case over user stories (also called scenarios). User stories are simplified descriptions of the user's requirements, and are based on the 3Cs: card, conversation, and conformation.

- *Card*: User story written as cards, and each user story has a short description about the story.

- *Conversation*: Requirements are gathered and refined through the continuous conversation between the users and development teams. Ideas and implementation design are acknowledged during the meeting with stakeholders.

- *Conformation*: Acceptance criteria of the user story are acknowledged during discussion about requirements with stakeholders. The user of the system confirms the conditions in which working software would be accepted or rejected.

User stories are easy to read, but a platform approach needs a more granular approach and detailed description about how the system will act. While doing use case analysis, we are designing a functional flow that meets the user's need. The digital platform-based approach is different than agile-based projects in terms of number of technology, framework, and systems involved in it; hence, behavior of the system should be analyzed in detail so that you are able to integrate and implement multiple technology, framework, and system efficiently and effectively.

In this chapter, you will examine the prospects of a DXP's requirements.

- Functional requirements
- Experience requirements
- Mobility requirements
- Security requirements
- Nonfunctional requirements
- Accessibility requirements
- Social and collaboration requirements

Let's begin with the functional requirements for a BXP.

Functional Requirements

The purpose of the functional requirements or functional specification document (FSD) is to understand the business requirement, develop a digital experience platform for an organization, and serve the needs of the client by workflow optimization and innovation of the business process.

An insight into the BXP use case of account and transaction in an ABC online banking portal is shown in Tables 2-1 and 2-2.

Table 2-1. Account Use Case Details

Use Case ID:	1
Use Case Name:	Accounts
Date Created:	XX-XX-XXXX **Last Revision Date:** XX-XX-XXXX
User goals:	To be able to view the account details
Primary actors:	1. Logged-in customer of the ABC bank
Secondary actors:	1. Master data management MDM) system 2. Core banking system (CBS)
Description:	Customer can view account details on dashboard on landing page, along with user profile data.
Trigger:	Customer wants account summary to be displayed on the screen. Customer should also be able to view details of account statement along with aggregated balance of all the accounts held.
Preconditions:	1. Customer should be an existing registered customer. 2. Customer will be logged in to ABC online banking application.
Postconditions:	1. All accounts of the customer are displayed on the screen.
Functional flow:	Following are the points to be considered for displaying account summary. **Account Summary** 1. Customer logs in with user ID and password. 2. Customer clicks on Accounts UI component on left-hand side of the landing page. 3. All accounts list will be displayed with the balance, along with aggregated available balance of saving/current accounts and outstanding balance of account and transaction details. • For example, savings account will show consolidated balance of all saving accounts along with individual savings account number, and current account will show consolidated balance of all current accounts along with individual current account number.

(continued)

Table 2-1. (*continued*)

Use Case ID:	1		
Use Case Name:	Accounts		
Date Created:	XX-XX-XXXX	Last Revision Date:	XX-XX-XXXX
Exceptional flow:	Backend system is not responding. 1. When back-end systems are not working, the UI component will show a human understandable error message to customer. 2. The error message that will be displayed is "**Sorry our systems are not working; please log in after some time.**"		
Assumptions:	NA		
Validations:	1. Account status validation: One holds many accounts in a bank; the system will display those accounts that have view permission, as per following table.		
	Value (Sent in web-service response)	**Validation**	
	OPEN	View	
	DORM	View	
	UNCL	Not applicable	
	INOP	View	
	CLOS	Not applicable	
	PRECREATED	Not applicable	
Track changes:	NA		
Out of scope:	NA		

Table 2-2. *Transaction Use Case Details*

Use Case ID:	2
Use Case Name:	Transactions
Date Created:	XX-XX-XXXX **Last Revision Date:** XX-XX-XXXX
User goals:	To be able to view the transaction history
Primary actors:	1. Logged-in customer of the ABC bank
Secondary actors:	1. Core banking system (CBS) 2. Simple mail transport protocol (SMTP)
Description:	Customer can view the ministatement. Customer can download or E-mail statement.
Trigger:	By default, customer wants ministatement displayed on the screen. Customer should be able to download statement on their machine.
Preconditions:	1. Customer should be an existing registered customer. 2. Customer will be logged in to the ABC online banking site. 3. Customer has selected the account to view statement and account details.
Postconditions:	1. Mini or detailed statement is displayed on the screen. 2. Detailed statement is downloaded in PDF format.
Functional flow:	Following are the points to be considered for transaction details: Account Statement 1. After login, customer will be landed on dashboard. 2. Dashboard will contain the transaction UI component (widget), which will display ministatement. 3. Ministatement will have details of last ten transactions. 4. Ministatement will be displayed on the screen with following details: date, expense type, description, credit or debited amount, and balance. 5. Customer can filter the transactions by selecting From date and To date to display detailed statement.

(*continued*)

Table 2-2. (*continued*)

Use Case ID:	2		
Use Case Name:	Transactions		
Date Created:	XX-XX-XXXX	Last Revision Date:	XX-XX-XXXX
Alternative Flow:	1. Customer can choose to send the statement via e-mail or download it in Excel or PDF format; respective buttons will be provided.		
Exceptional Flow:	1. If there is no transaction in the account, appropriate message will be displayed: "**Transaction details are not available for this account.**" 2. Backend system and web-service calls are not responding. 3. Error message will be displayed: "**Sorry our systems are not working; please log in after some time.**"		
Assumptions:	NA		
Validations:	1. From date cannot be prior to account opening date and To date cannot be greater than today's date. 2. From date and To date period is six months.		
Business rules:	1. Maximum period to display or download or mail a statement is six months at a time.		
Track changes:	NA		
Out of scope:	NA		

The use case shown in Tables 2-1 and 2-2 should help you understand how the BXP application interacts with multiple systems and platforms to provide an omnichannel experience to the bank's customers. The BXP interacts with the MDM system, core banking system, and SMTP mail server and provides cross-channel capabilities where the user can log in to the BXP from a mobile device or desktop and get the same experiences across all platforms.

Experience Requirements

Let's look at experience requirement for a BXP. The purpose of the experience requirements is to understand workflow across all channels, to provide a seamless and smooth user experience to the customer. Experience requirement help you to choose a technology stack for the DXP. The user story is the best option while gathering experience requirements, because the experience specifications of the application are explained from a business point of view. Business and management will explain the specific features and channels that they want to provide to their users.

The DXP is designed for the entire user journey. The DXP provides a better user experience with your organization through multiple channels. You should understand when and why users move across channels, which further helps you to design efficient and smooth user experiences. The DXP should provide zero or minimal overhead while transiting from one channel to another.

Seamless Experience on All Supported devices

The user interacts with the application from multiple channels, that is, Web, mobile device, tablet, e-mail, chatbot, and IVR, etc. (as shown in Table 2-3). Experiencing failure on any channel reflects a bad experience as a whole. You should consider the workflow by considering all channels supported by your organization. You need to consider touch points requirements across all channels. For example, consider a scenario where a user makes a service request using a web application on a desktop, but is unable to complete the process or is confined due to any personal reason. The user can still continue the process on their mobile device. If they can pick up from where they left off, their user experience will be seamless.

Table 2-3. *Device Support User Story*

Name	Device Support
Trigger	Customer wishes to access a banking portal from different devices, that is, mobile, tablet, or desktop.
Script	A customer can access the banking portal application on a mobile device, tablet, or desktop. A person will be able to continue an action that is left off in a previous session, irrespective of the device that person was working on.
Acceptance criteria	The customer will be able to access the BXP application on a mobile device, tablet, or desktop.

Seamless Experience on All Supported Browsers

The user can interact with your application from multiple web browsers. Therefore, once the application is developed, it is supported across all the latest browsers: that is, IE 10 and above, Chrome 2.0 and above, Firefox, etc. See Table 2-4.

Table 2-4. *Browsers Support User Story*

Name	Browsers Support
Trigger	Customers can log in to the application from different browsers.
Script	A customer can log in to the BXP application through the following browsers: • IE 10 and above • Chrome • Firefox
Acceptance criteria	The customer will be able to access the banking portal on IE, Chrome, or Firefox.

Multilingual Requirements

You need to consider language criteria on the basis of business requirement while developing a DXP application. A DXP provides the capability to store language preference in user preference. On the basis of language preference, content will be shown to the user. See Table 2-5.

Table 2-5. *Language Support User Story*

Name	Language Support
Trigger	The customer can select the language of the content.
Script	A customer can select the preferred language while registering into the BXP portal. One can select from the following languages: • English • Hindi • German • French After login to the BXP portal, one would get the content according to one's preferred language.
Acceptance criteria	The customer is able to get content on the basis of preferred language.

Navigation Elements, Menus, and Search

An experience is either a website or a mobile application. Each of these experiences consists of pages. A page consists of containers and UI components. The visible elements of an experience are the pages, the containers, and the UI component. Pages have containers and UI component as children. Containers can have other containers or UI component as children. A page has an associated URL, stored in a link. The UI component displays the content or functionality. Containers group UI components together in a visual layout. We will go through these components in detail in Chapter 4.

Navigational routers and elements help to navigate from one location to another, to enhance the user experience. Navigation routers are able to provide menus in an efficient and interesting way, and search capability enhances the accessibility features. DXP search capability helps you to find the intent of the user, to provide the most precise result. These components help to provide information in an appealing as well as an organized way. An example is a banking dashboard, where the user is able to get frequently visited UI components, that is, user details, accounts, and transactions related to a specific account on one single page.

You should consider the user experience journey before developing UI components. Every user story is bifurcated into individual functionality, and these individual functionalities are considered to be one UI component. For example, in the case of the dashboard user story in Table 2-6, Account is considered one functionality and Transaction is considered another functionality; that means creating two UI components: one for accounts and another for transactions.

Table 2-6. Dashboard User Story

Name	ABC Bank's Dashboard
Trigger	As a customer, one wishes to view Account Summary and Transaction Statement on one's dashboard after login to the banking portal.
Script	As a customer, after login to the portal one can see a navigational plane on the left hand-side and component plane on the right-hand side. As a customer, one can navigate to another page through the navigational router plane. Account and Transaction are presented in the component plane on the right-hand side.
Acceptance criteria	The customer is able to get frequently visited UI components, that is, user details, accounts, and transactions related to a specific account on one single page.

CHAPTER 2 GATHERING REQUIREMENTS

The navigational router will be the dashboard, whereas user details, account, and transaction will be different UI components in multiple containers, as you see in Figures 2-4 and 2-5. Searching and filtering transactions enhance accessibility.

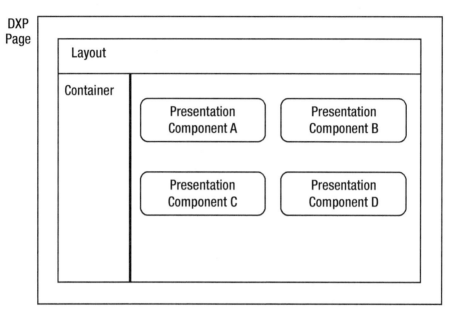

Figure 2-4. User interface components

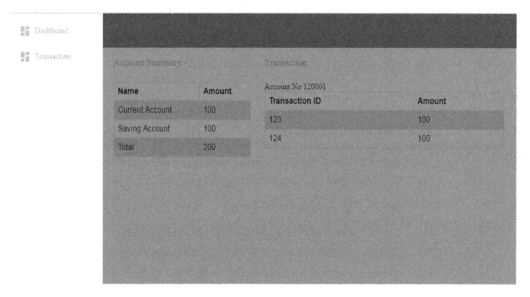

Figure 2-5. BXP dashboard

We will go through the BXP dashboard in detail in Chapter 4.

Mobility Requirements

Nowadays you should be future ready. BXPs support native applications, that is, a smartphone application developed specifically for a mobile operating system as well as hybrid applications (i.e., a hybrid application at its core is websites packed into a native wrapper). You have to consider mobility requirements while developing the DXP application.

Table 2-7. *Mobility User Story*

Name	Mobility
Trigger	The customer should be able to access the BXP application on Android and iOS platforms along with web support.
Script	As a customer, one will be able to download a BXP mobile application from iOS App-Store and Android Play Store.
Acceptance criteria	The customer will be able to download the BXP application to a mobile device and be able to access all functionality provided on a web application through the mobile application.

The DXP supports responsive, native as well as hybrid applications approach. While designing a responsive application, you need to consider the mobile-first approach. From a general point of view, mobile designing is the hardest as compared with other devices, in view of the small screen to which you can provide essential features. In an opposite approach, if you design the all-inclusive right from the start, the core and supplementary elements merge and it will become difficult to distinguish and separate. Mobile first is equivalent to content first, due to limited screen size and bandwidth; therefore you should prioritize content. A responsive web approach would have a lack of interaction with mobile features like sensors, camera, etc. Native applications and hybrid applications provide additional features to interact with device hardware, for example, fingerprint readers, camera, etc. Consider a scenario where you want to give a notification feature to your user: a native application or hybrid application is configured with push notifications features, so whenever a user does a transaction, these applications push notification to the user's mobile device. A hybrid application has the advantage of a single code base, which makes it compatible with all browsers and devices, whereas a native application has the advantage of providing a highly interactive and rich experience, and exploiting the entire native device features such as sensors and camera, etc., and provides high performance over responsive and hybrid applications.

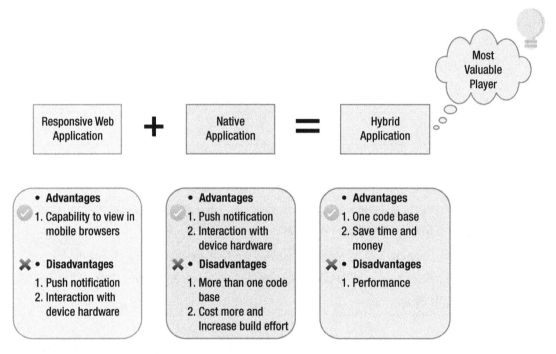

Figure 2-6. *Responsive vs. native vs. hybrid*

You need to go through the usage of the application because a hybrid application is a web application (shown in Figure 2-7), built using HTML 5 and JavaScript's wrapped-in native container, which loads most of the information on the page as the user navigates through the application, whereas native applications instead download most of the content when the user first installs the application.

CHAPTER 2 GATHERING REQUIREMENTS

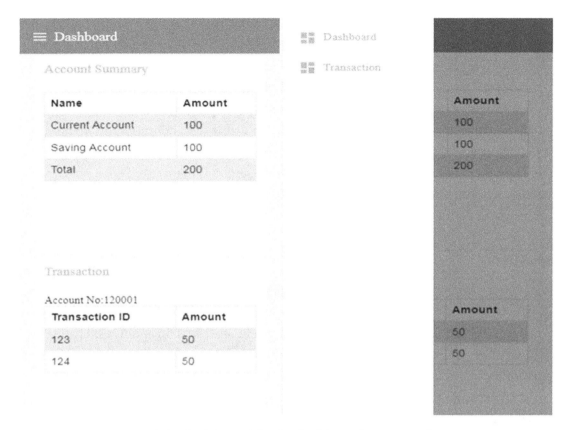

Figure 2-7. *BXP mobile dashboard (Left: dashboard view; Right: navigation view)*

The native application has the best performance and highest security. The performance of the application as well as the user experience vary significantly, based on the development framework chosen, along with the native application approach, the overall performance, and security improves.

Nonfunctional Requirements

Nonfunctional requirements (NFRs)—also known as quality attributes—decide the robustness and long-term success of the DXP. The quality attributes such as scalability, usability, reliability, availability, maintainability, and performance are the key NFRs that help us to define, track, and measure the success metrics of the digital platform.

There are other NFRs also, like serviceability, security, regulatory, environmental, data integrity, usability, interoperability, etc., but RAM (reliability, availability, and maintainability) is most pertinent to a DXP. These requirements help to understand the operations of the system rather than specific behavior.

Consider a scenario where you have created a web application to have an eye-catching and adaptive UI design. But what if it is not able to handle appropriate traffic? A digital experience platform ensures that balance between utility of the service (functionality) delivered and warranty, that is, whether it is fit for use. The perfect balance between functionality and its use creates maximum value to customer.

Scalability Requirements

Scalability requirements ensure maximum operating capacity of an application and determine whether the current infrastructure is sufficient to run the application. This provides a holistic view of the number of concurrent users that an application can support, and ensures scalability so that application can support and allow more users to access than its current operating capacity.

It is necessary to look into the scalability requirements, as mentioned in the scalability user story in Table 2-8.

Table 2-8. *Scalability User Story*

Name	Scalability
Trigger	The DXP application should be scalable and load should be distributed across geographical locations.
Script	As a product owner, the DXP application will support 10,000 concurrent users per hour and 1,000,000 transactions per hour. The application should be robust so that it will be able to handle a heavy request load. Geographical load should be distributed across locations so that the application will be available across all geographical locations.
Acceptance criteria	The DXP application is able to support 10,000 concurrent users with 1,000,000 transactions per hour across all geographical locations.

CHAPTER 2 GATHERING REQUIREMENTS

When you need to determine the number of real simultaneous users that the application can support, you first need to calculate the maximum throughput.

Maximum throughput is calculated by running a few emulated users with zero think time. That means each user sends a request, receives a response, and immediately loops back to send the next request.

- *Maximum users the application has to support*: Once you have the maximum throughput, you can use Little's Law to estimate the number of maximum concurrent users that the application can support.

    $N = X / \lambda$

    N is the number of concurrent users.

    λ is the average arrival rate.

    X is the throughput.

- *Maximum concurrent users*: If the application has ab average interarrival time of 5 seconds, using Little's Law we can now compute N (number of users) as:

    $N = X / \lambda = 2015 * 5 = 10075$ users.

    Your application running on the same infrastructure can support more than 10,000 concurrent users with an interarrival time of 5 seconds.

- *Maximum concurrent volume*: You need to calculate maximum concurrent users per hour that need to be supported by the application.

- After estimating the concurrent user and load supported by the DXP application on the particular infrastructure, you can decide to scale your infrastructure accordingly.

- *User growth rate*: You need to estimate percentage increase in user traffic per year so that you are able to scale up your infrastructure in the future.

- *Content growth rate*: You need to estimate percentage increase in content volume per year; that will help you to analyze your load capacity per year.

- *Average session time*: It is the average amount of session time for a user, and helps you to analyze and estimate in-memory support required in the near future.

- *Geographic (Geo)-specific load*: Globally available applications should distribute load across geographical locations. Performance, availability, and scalability should be specified for each geographical location.

- *Peak volume or traffic time*: The maximum amount of users that should be supported by the application at peak business hours.

- *Data volume*: As data volume keeps changing and it depends upon load and usage of the application, you should estimate the average amount of data that should be handled by the DXP application, which will let you estimate the disk space requirement in the near future.

Performance–Response Time, Throughput, Utilization, Static Volumetric

It is essential to check your industry standards for measuring application performance. Results should be collected from real browsers, which will assist you in checking the page load time on different browsers and operating systems.

Gain insight into the performance requirements and testing approaches. Before beginning with load testing, you need to determine page response time applicable to the business process and whether response time is justifiable and achievable.

Performance Requirements

You need to consider and build workload profiles for the user stories as mentioned in Table 2-9, and associated workload profile related to use case mentioned in Table 2-10.

CHAPTER 2 GATHERING REQUIREMENTS

Table 2-9. *Performance User Story*

Name	Performance
Trigger	The DXP application should be able to respond within 3 seconds.
Script	As a product owner, the DXP application will respond to each and every customer request made through the customer's browser within 3 seconds.
Acceptance criteria	The DXP application is be able to respond within 3 seconds.

Table 2-10. *Workload Profile*

FSD's Use Case	Daily Total	Pages	Time
Account	20,000	Login, Dashboard (navigational router and account component).	3 Sec.
Transactions	15,000	Login, Dashboard (navigational router and transaction component)	3 Sec.

Once all loads have been considered, then infrequent or inappropriate workloads can be eliminated.

Page Response Time at Normal and Peak Loads

In supporting 10,000 users, the DXP should certify that performance should not fall below the mentioned level.

- 80% of all pages for customers respond in 3 seconds or less.
- Transaction and Account pages should respond in 3 seconds or less.

In Figure 2-8 and Table 2-11, consider the following:

- *Peak load*: When the maximum number of users engages with your application, as shown in Figure 2-8: Peak Load
- *Normal load*: When frequent users engage with your application, as shown in Figure 2-8: Normal Load

47

Table 2-11. *Peak Load and Normal Load Analysis*

Page/Response time.	Peak Load (Seconds)	Normal Load (Seconds)
Transaction	3	2.3
Account	3	2
Dashboard	6	4

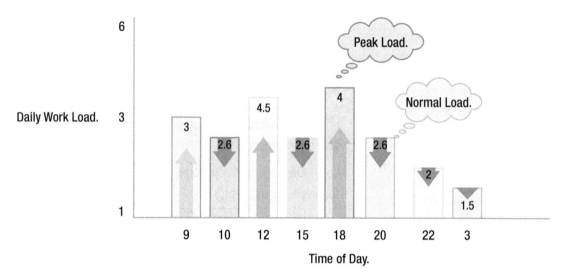

Figure 2-8. *Normal load and peak load*

When defining the workload for a new application and no existing workload data exists while the system is being developed, the temptation is to specify a high peak workload.

Page Hits Analysis

You need to consider page response time at normal and peak load for various geographical locations, where you are providing business to your users. This will help you to understand and divide the loads according to the available infrastructure on the basis of workload profile.

- Page response time at normal and peak loads for (Hypertext Transfer Protocol) HTTP or (Hypertext Transfer Protocol Secure) HTTPs pages

CHAPTER 2 GATHERING REQUIREMENTS

Pages load = resource download + service calls

- Where resource download is average time taken to download resources like Cascading Style Sheets (CSS), Hypertext Markup Language (HTML), scripts, and images, etc.; and service call is average time required by the web services to return data from the server.

- You need to build a service call workload profile for all the services calls impacting the particular page.

• *Transaction time*: The average time taken for key transactions, for example, average time taken by account and transaction services to fetch the data as well as other resources (e.g., CSS, HTML, scripts, and images).

• *Search completion time*: The average time taken by the search module to provide the top ten results. *Performance Testing*: The DXP application must fit performance levels and delivery times on the agreed SLA (service level agreement) for stress testing to be planned and performed.

- Load testing and stress testing will be performed on the BXP portal pages to certify that the critical performance requirements are met.

- The pages will be performance tuned to guarantee that the response time is within 2 to 4 seconds for all the pages under average production load.

- The target CPU utilization will be under 25%.

- The LoadRunner application will be used to perform load testing.

Maintenance Requirements

You should be bringing its rich experience in successful execution of large-scale portal engagements in continuous application or system SLA monitoring and maintenance.

You should be building robust monitoring applications to confirm that the application maintains the SLA related to performance, availability, and scalability:

- Real-time application SLA monitoring components check the live production BXP web pages. The frequency of page URLs can be configured.

- Automatic alerts and notification through e-mail or page when the page or system performance falls below a preconfigured threshold value.

- System health-check or heartbeat monitoring to ping the availability of the portal system and all interfacing systems, to ensure that they are responding within good response time. Automatic notification gets trigged if any system is down.

- Web analytics will be configured to monitor the business-critical process or activities in real time. This includes activities such as page load time, search processing time, etc. Additionally, reports will be designed to display the monitoring data, based on requirements.

CHAPTER 2 GATHERING REQUIREMENTS

Table 2-12. *Monitoring User Story*

Name	Monitoring
Trigger	The DXP application should be monitored.
Script	As a product owner, I want real time monitoring of the application so that if the system fails, automatic alerts and notification will be sent through e-mail, which ensures 24/7 availability of the application. Business critical process should be monitored 24/7.
Acceptance criteria	The DXP application is monitored.

Table 2-13. *Serviceability User Story*

Name	Serviceability
Trigger	The DXP application should be serviced.
Script	As a Product owner, I want to service the DXP application for system cleanup. Maintenance activity should be scheduled so that back jobs perform system cleanup and take backup.
Acceptance criteria	The DXP application will be serviced and regular backup maintained.

Table 2-14. *Maintenance User Story*

Name	Maintenance
Trigger	The DXP application should be maintained through versioning.
Script	As a Product owner, I want to maintain the DXP application through a version control system so that incremental changes are released to the production environment. Bug fixes should be published thought a rollout and change request mechanism to ensure smooth and effective rollout of the bug fix release to the production environment.
Acceptance criteria	The DXP application is maintained through versioning.

Chapter 2 Gathering Requirements

Versioning

For Source Code Management (SCM), we can use versioning tools such as CVS, MS Source Safe, Git, and SVN for versioning control of the application.

Rollout

Once the system is developed and moved to the production environment, rollout of new functionality goes through rollout protocols and procedures. These aid in analyzing the changes involved in moving to production, and help ensure smooth and effective rollout of new functionality.

- Rollout of any new functionality and bug fixes goes through the change management process after assessing the impact to cost and schedule.

- Rollout includes rollout release name, rollout details, device support, release history, and defect reports.

    - *Release name*: Name to identify release, such as an internal code name or build version

    - *Rollout details*: A timestamp indicating the last rollout event for each release

    - *Device support*: A summary of the application device compatibility, including supported devices

    - *Release history*: A list of all previous releases with version code details, rollout history, and release notes, which contains all of the aforementioned details along with defect or testing reports.

- *Defect management and reports*: You should follow a proven defect management process for a BXP. It has the ability to tailor the process to align with the current process or objectives. The key elements of the defect management process involve the following areas:

  - Set up and customize defect tool
  - Define defect lifecycle flow
  - Define and publish defect classification
  - Identify key stakeholders and their responsibilities
  - Go through defect triage process
  - Manage defect resolution
  - Report and escalate procedures
  - Conduct root cause analysis of defects
  - Defect classification and turnaround time will be decided during the test strategy phase.

Security Requirements

Security is a main concern while developing an application. You ought to consider all the aspects revolving around security. DXP security enhances the user experience and ensure data integrity, authenticity, and authorization.

A banking experience platform is always vulnerable to attacks. You need to go through the requirements while designing a banking portal.

CHAPTER 2　GATHERING REQUIREMENTS

Table 2-15. *Session Management Use Case*

Use Case ID:	3
Use Case Name:	**Session Management**
Date Created:	XX-XX-XXXX　　　　　　　　　　**Last Revision Date:**　　XX-XX-XXXX
User Goals:	1. To log in to online banking portal of ABC Bank
	2. To log out from online banking portal of ABC Bank
Primary Actors:	1. Existing customer of the bank
	2. E-banking application – BXP
Trigger:	1. Logout from online banking portal after clicking on logout button or automatic session timeout after defined time limit
Preconditions:	1. Customer of ABC Bank is registered with Internet banking portal.
	2. Customer logs into Internet banking portal of ABC Bank.
Postconditions:	1. System will be logged out from the online banking portal of ABC Bank.
Functional Flow:	1. Customer logs into ABC online banking portal.
	2. Customer will be landed to the Dashboard.
	3. In case customer wants to log out from online banking portal, customer should click on logout option on landing page at any point in time.
	4. Customer will be logged out from the present page, and home page will be displayed on the screen.
	5. After 5 minutes of idle time, portal will invalidate a session and login page will be displayed on screen.
Exceptional Flow:	1. When backend systems are not working, UI component will show human understandable error message to customer: "Sorry our systems are not working; please log in after some time."
Assumptions:	NA
Validations:	Session timeout
	After 5 minutes of idle time, portal will invalidate a session, and session timeout is configurable.
Track Changes:	NA
Out of Scope:	NA

CHAPTER 2 GATHERING REQUIREMENTS

- *Session Management Considerations*:

    a. It depends upon the business requirements of the user whether you want in-memory token session management or database token session management (also called Java Database Connectivity [JDBC] token management). JDBC token management is helpful in case of clustering the BXP application; in in-memory token session management, you need to replicate the session token across all clustered environment using JGroups or other open-source cache replicating techniques.

    b. You need to decide the idle time of a session on the basis of business requirements. Idle time of a session will automatically log out the user from the server after a specified time.

It also requires checking authenticity, authorization, and integrity of data while designing the UI layer as well as backend integrations.

Look into the BXP authenticity and authorization user story in the ABC online banking portal (Table 2-16).

Table 2-16. Authenticity and Authorization User Story

Name	**Authenticity and Authorization**
Trigger	Customer should be authenticated and authorized.
Script	As a customer, one would be logged in using two-factor authentication, so that authorized data for that customer should be accessible. The following items should be masked: • Mobile device number • SSN • Account number
Acceptance criteria	An authorized and authenticated customer is able to access the BXP application.

CHAPTER 2 GATHERING REQUIREMENTS

- *Authentication Consideration*:

  a. You should always opt for two-factor authentication. There are multiple two-factor authentications.

     - Username / Password + one-time password (OTP).

     - Username / Password + RSA public key cryptography algorithm-based questions and answers.

     - *Both*: In this case, if the system finds an irregularity or unusual pattern during login, the system can trigger OTP and RSA questions after login; this eliminates vulnerability and provides high security.

  b. *Number and password masking*: You should mask the password and numbers, for example, mobile device number and account number.

  c. *Cross-site request forgery (CSRF)*: is an attack that forces an end user to execute unwanted actions on a web application in which they are currently authenticated. You need to implement Open Web Application Security Project (OWASP) CSRFGuard; that would protect the application from CSRF attack.

Table 2-17. Integrity User Story

Name	Integrity
Trigger	DXP application should provide data integrity while retaining data interoperability between client and server.
Script	As a product owner, I want to maintain the integrity of data while retaining data interoperability between client and server, so that the communication channel will be secured. Tokens should be used and passed with data for transmitting information between client and server.
Acceptance Criteria	DXP application provides data integrity while retaining data interoperability between client and server.

- *Service calls—data interoperability*:

    a. JSON Web Tokens (JWTs) are used for transmitting information between parties as JSON objects. This information can be secured by using a secret key using a hash-based message authentication code (HMAC algorithms or a public or private key pair using RSA algorithms). JWTs ensure the integrity of data transferred as well.

    b. Cross-site scripting (XSS) filters: You use the validation library to verify web service application programming interface (API) requests in line with (Java Specification Requests) JSR standards. XSS filters match suspicious content in data requests and reject them if there are matches.

Disaster Recovery Requirements

Disaster can be any situation that makes an organization's operations prone to risk; it can be of any type, for example, natural disaster, equipment failure, or cyberattacks. Disaster recovery (DR) requirements help to continue business operations as normally as possible. You need to find the recovery point objective (RPO) and recovery time objective (RTO) for their DXP application. The RPO is the maximum duration of an application (age of Files, Database, User Sessions and Caches) that an organization must recover from backup for normal operation to resume after a disaster: for example, a DXP application has an RPO of 2 hours, and then the system must back up at least every 2 hours. The RTO is the maximum duration of time for an organization to recover an application from backup storage: for example, if the organization has an RTO of 1 hour, it will not be down for longer than that. Table 2-18 contains the disaster recovery requirements.

Table 2-18. Disaster Recovery User Story

Name	Disaster Recovery
Trigger	The DXP application should be able to recover in case of any disaster.
Script	As a Product owner, I want the DXP application to have an RPO of 2 hours, so that the application will back up every 2 hours and will reinstate the application to the backup point.
Acceptance Criteria	The DXP application is recovered in 1 hour.

RTO and RPO help you make disaster recovery strategies.

- Identify the threats related to your application.
- Identify relevant infrastructure documents, for example, utility diagrams, HVAC diagrams, network diagrams, and equipment configurations.

Accessibility Consideration

You can boost the DXP capabilities by including accessibility best practice that removes barriers that prevent interaction with a web application. You need to consider the different aspects of accessibility requirements. Following is a list of factors that help your business use case to ensure web accessibility. Including these requirements while preparing a use case will improve search engine optimization (SEO), interoperability and quality, and reduce web application development, maintenance time, effort, and server load, etc.

- Provide equivalent alternatives to auditory and visual content.
- Provide clear navigation mechanisms.
- Create tables that transform gracefully.
- All functionality will be available from the keyboard.
- Text content will be readable and understandable.
- Provide enough time for the user to read.
- Content must be robust enough, as it will be accessed and interpreted by a wide variety of devices and technology.

You must check and consider the aforementioned points before developing an application, or else it will be difficult for one or more than one group to access the web content.

CHAPTER 2 GATHERING REQUIREMENTS

Chapter Summary

- This chapter covered almost every aspect of DXP applications requirements: functional, experience, mobility, security, nonfunctional, and accessibility requirements.

- Experience requirements cover the following user stories along with their considerations.

    - Device support user story

    - Language support user story

    - Dashboard user story

- Security requirements cover the following use case and user stories along with their considerations.

    - Session management use case

    - Authenticity and authorization user story

    - Integrity user story

- Maintenances requirements cover the following user stories along with their considerations.

    - Monitoring user story

    - Serviceability user story

    - Maintenance user story

- Banking experience platform requirements have been covered, which in turn should aid you to go through your own requirements before developing a DXP application.

CHAPTER 3

Design

After going through the DXP requirements in Chapter 2, we will now look into designing and building a Digital Experience Platform in detail. As the world moves toward a modern digital economy, the use of DXP is meant to provide an ecosystem for product and service innovations, in addition to providing a space for the organization's activities.

Building an Experience Platform

We will look at the design of the following layers in detail throughout this chapter.

- Presentation layer
- Business layer
- Integration layer
- Data layer
- Middleware layer

We will also look into integration of the following cutting-edge digital technology.

- Social and collaboration design
- IoT integration design
- Blockchain design
- AI automation design
- Big data and NoSQL design
- Enterprise search engine design
- Augmented reality design
- Recent trends in DevOps

CHAPTER 3 DESIGN

DXP design principle works on creating brand value, using visual design, interaction design, along with information architecture, as shown in Figure 3-1.

Four principles for designing a digital experience platform are as follows:

- *Brand Value*: Social platforms help your organization to promote and position your brand to social media channels like Facebook, Instagram, LinkedIn, etc. Social platforms have fast gained adoption, considering there is a strong social incentive to use them. When people are utilizing an application to reach out to you, then you should build social networks. For that, you need to choose a solution that incorporates microblogging, social networking, dynamic profiles, and automated activity feeds. You need to decide on many factors on which social software solutions will be integrated with DXP. You can integrate wiki for information sharing, Facebook authentication, and social feed APIs to integrate user social interaction with DXP that will further enhance the user experience so that one can interact with your brand across every digital touch point.

- *Interaction Design*: Interaction design is design where one defines the structure and behavior of interactive systems. DXP user interface (UI) design helps you to create an interface design in such a way that makes the state of the underlying system easy to use and understand. User behavior is carefully examined, which ensures smooth navigational and application usability design along with seamless workflow across all channels and touch points.

- *Visual Design*: DXP's focus on visual helps you to build an elegant and jazzy UI using the latest Material CSS Design approach. This principle ensures the proper usage of imagery, color, typography, and form to enhance usability and improve the user experience.

- *Information architecture*: DXP enables you to use information architecture (IA) to get insight from every customer interaction on every touch point such as desktop applications, Internet of things (IoT) devices, interactive voice response (IVR), and mobile applications into a single customer view. The IA principle is to ensure that data analytics and continuous learning and improvement approaches will help you understand the data generated through

CHAPTER 3 DESIGN

these touch points. Data acquired from these touch points are used to do prediction followed by classification, detection, automation and recommendation systems, using AI/ML concepts and algorithms.

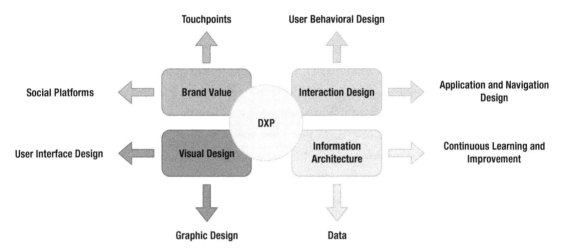

Figure 3-1. *DXP customer-centric design principle*

DXP has a six-layered approach, as shown in Figure 3-2, to achieve the four principles shown in Figure 3-1. That is to achieve a seamless experience across all digital touch points using the platform approach (touch points, UI, integration, continuous learning and improvement, data analytics and data delivery, and infrastructure).

- *Touch points*: Touch points are also called interaction channels. Each organization needs to understand their own digital strategy, as this will help to understand their touch point needs and create brand value.

- *User interface*: Interactive and intuitive UI help your organization enhance the usability of your products and services.

- *Continuous learning and improvement*: Continuous learning helps you to improve their business process and optimize business workflow, for example, chatbot or AI automation using machine learning (ML) algorithms and neural networks help you to automate and optimize their business process. DXP is designed to understand the context through transaction and learning engines where you can use different ML algorithms to solve problems related to prediction, time-series analysis, recommendation, etc. You can design the experience platform by continuous learning of user experience and understanding the intent of your customer, through which you can innovate their services and products.

- *API ecosystem (integration)*: The API ecosystem is playing a critical role in creating digital business and a huge digital economy. API helps you to turn a business and organization into a platform. API is game changing while integrating and connecting systems, data, IoT, and algorithms. In the present world, where the economy runs by API ecosystems, for example, Uber used Google API to build a new business platform; while buying a movie ticket online, one uses API to verify a customer's debit or credit card information. The API ecosystem is playing a vital role in enabling emerging technologies like AI, blockchain, and IoT to connect with each other and build smart applications and devices.

- *Intuitive data*: Data helps you to get insight and recognize patterns, and display reports on the basis of latest trends that are used for predictive analysis, user behavior analytics, and advanced data analytics that extract value from a particular size of data sets. For example, you can use an open-source NoSQL database that gives better performance in storing a huge amount of data and provides analysis and reporting features as well. A DXP is also designed with the consideration of customer experience management (CXM) and customer relationship management (CRM). DXPs get insight about what a customer thinks about a brand, and help an organization to know about their customers. We will get insight about interaction through touch points, further tracking of pages, and UI components, so that the organization is able to track user experience to provide a better experience to the customer.

- *Fast delivery (infrastructure)*: Infrastructure depends upon many factors, whether one wants a hassle free and application-focused platform, such as cloud infrastructure or wants their own data centers and virtualized server. Cloud infrastructure provide faster delivery capabilities. The application should be cloud ready, as cloud-based technology will help to grow the infrastructure quickly.

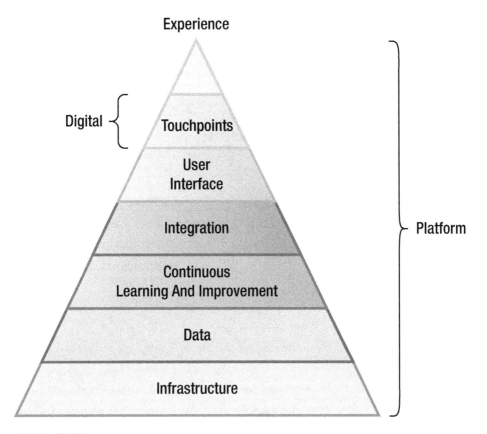

Figure 3-2. DXP

DXP key focus areas are platform principles that enhance the following key areas:

- Reusability of components
- Extensibility of components
- Robustness and scalability of platform
- Quality focus on performance, security, and availability

Digital Platform Strategy

Digital platform strategy is to design an ecosystem that gains insight, delivers service faster, and provides consistent experience across all channels. A DXP-based organization has an innate innovation capability. The platform approach helps to provide limitless digital capabilities along with providing a more innovative approach to solving issues related to additional debt and cost such as inventory, staff, billing, and management. Table 3-1 will help you to establish the relationship between DXP design principle and its strategy and platform approach to achieve it.

CHAPTER 3   DESIGN

Table 3-1. *Digital Platform principle and Strategy*

Digital Experience Platform	Design Principle	Digital Platform Strategy
1. Touch points, for example, desktop, mobile, tablets, smart devices like Amazon fire stick, Alexa, etc. • User Interface	• Brand value • Visual design • Interaction design	• Selection of touch points. • Workflow across touch points so that user gets a seamless experience across all devices.
2. Continuous learning and Improvement • Selection of ML algorithms • Selection of neural networks like RNN, ANN, etc. • Selection of artificial intelligence (AI) API like tensorflow, pytorch, etc.	• Information Architecture	• Continuous Improvement, which includes usage and implementation of a chatbot, prediction analysis, time series analysis, etc.
3. Data. • Database Design • Database selection	• Information Architecture	• Intuitive data includes pattern analysis, data storage strategy, that is, choosing NoSQL and SQL databases, and designing data scalability approach.
4. Integration • API gateway • ESB approach • Microservices or monolithic approach	• Interaction design	• API ecosystem includes strategy to choose microservices approach or monolithic services approach, choosing ESB and gateways for API integration.
• Infrastructure	• Information Architecture	• Fast delivery infrastructure, which includes platform security scalability and deployment approach.

Digital platform strategy, as provided in Table 3-1 helps one to build their own DXP for their organization. Let's look into mapping of these principles to DXP strategy and design as shown in Figure 3-3.

- *Touch point selection and design*: You select the touch point that helps you to provide interactive services through channels such as IoT devices, Web, and mobile, as shown in Figure 3-3. Augmented reality (AR), virtual reality (VR), and IVR are by-products built on top of these touch points to make your application intelligent and interactive.

- *Integration (API ecosystem) design*: Integration of data from the different services (such as web API), databases (transactional and nontransactional), devices (data collected from IoT devices and smart devices through data pipeline), and systems (such as a blockchain ecosystem) help to build an integration ecosystem using an enterprise service bus (ESB) and API gateways so that multiple systems, devices, and applications connect and transfer information irrespective of technology, protocols, and frameworks used.

- *Continuous learning and improvement design*: Continuous learning and improvement helps you to include AI and ML capability in your organization, as shown in Figure 3-3. Natural language processing (NLP) provides capabilities so that you can build efficient search engines for the organization; chatbot to deal with repetitive problems faced by user in interactive way; and if-else analysis to predict, train, and test ML and AI models built for the organization to solve day-to-day problems. Across industries, one can turn their application to AI/ML-driven smart applications that help to drive automation using devices such as IoT devices, mobile phones, etc. We will look into the libraries and frameworks used to develop it in AI automation design and Big data and NoSQL design in this chapter.

- *Data design*: Data pipelines, and database design and approaches help you to provide big data analysis capability to your business. AI-ML is so powerful in terms of what it can do but it needs tons of data to learn, hence to manage the data in real-time you should use a distributed data approach, hence data pipelines are introduced.

- *Infrastructure design*: Digital strategies need to engage users with a high-performing experience regardless of the location where it is deployed. You can use cloud infrastructure or standalone server

CHAPTER 3 DESIGN

infrastructure, according to the application design and development approach you take while building the DXP. The cloud and container architectures power the DXP's application; we will look at containerization in brief in the Containerization section.

- *DevOps*: DevOps (development operations) such as continuous development, integration, deployment and delivery help you to develop and deliver faster on a DXP. As shown in Figure 3-3, an application incorporating continuous learning strategy, intuitive data strategy, and API integration strategy is built and deployed using an agile approach using DevOps.

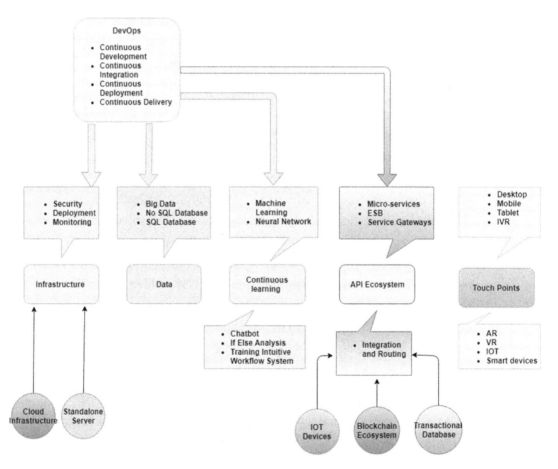

Figure 3-3. *Digital platform strategy*

Platform Design Phases

Designing a platform involves software development life cycle (SDLC) phases. It ensures quality, reduces cost, and saves time. SDLC phases are essential for the success of organizational DXP management strategy. Five phases are: explore and elaborate, design, prototype, validation, and delivery, as shown in Figure 3-4. You explore requirements and elaborate those requirements, build your design and on the bases of your design build your prototype, validate the prototype, and reiterate the process until the business requirements are fulfilled, and then deliver the application after ensuring quality and testing.

- *Explore and elaborate requirements*: Investment and DXP strategy cannot be estimated accurately until the requirements are clear. You need to understand the intention and problem statement, as we observed in Chapter 2 in detail.

- *Design*: On the basis of digital requirements, you will work on strategic designs. We will look into strategic design for application development in detail in this chapter. Keep all stakeholders engaged in the design process.

- *Prototype*: Model the requirement. After creating strategic designs, you build prototypes, which helps your organization to make new innovations in products and services with optimized processes.

- *Validation*: Always conduct acceptance testing. Once a prototype is built, it is validated on the basis of design: whether the prototype is appropriate according to requirements or not.

- *Delivery*: Once the prototype is validated according to functional requirements as well as nonfunctional requirements (NFRs), test scenarios are tested.

After passing unit testing, integration testing, user acceptance testing, and preproduction tests, it will be moved to production.

Chapter 3 Design

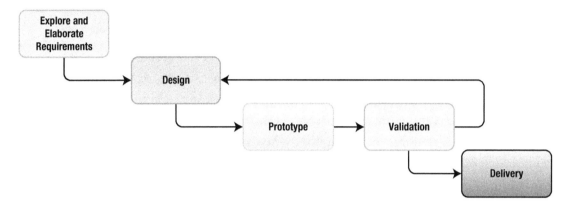

Figure 3-4. *Platform design phases*

Design of Various Layers

A DXP is a flexible platform that has the capabilities of rapid development and innovations. A DXP has a web application with a content management system (CMS) and provides features such as marketing, targeting, personalization, commerce, etc. You can build web applications, mobile applications, chatbots, AR, VR, AI, and enterprise search engine components for devices such as tablets, mobile, desktop, and IoT, as shown in the presentation layer in Figure 3-5. These applications on different devices are connected through the integration layer; you can choose architecture type such as microservices or monolithic application, and services built are exposed using an ESB and API gateway. Business logic is implemented while accessing data from the data access layer and exposed using the integration layer, as shown in the figure. The middleware layer is responsible for implementing NFRs such as reliability, security, accessibility, and scalability, etc.

CHAPTER 3 DESIGN

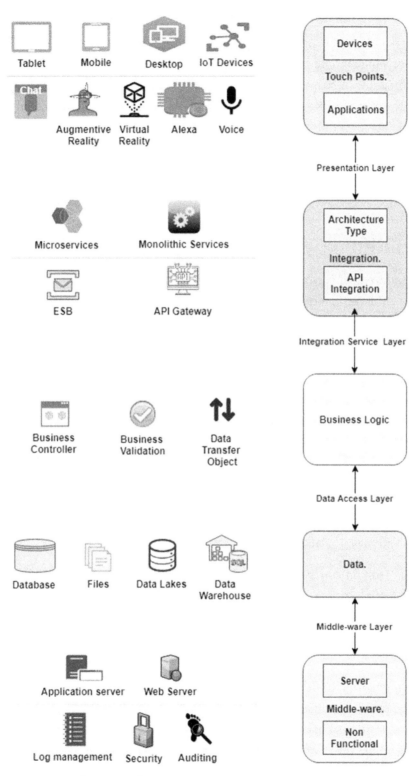

Figure 3-5. *Design of various layers*

71

CHAPTER 3 DESIGN

Each layer is important while designing a DXP's application. The presentation layer helps you to decide user experience on user interaction points (touch points). The service layer is responsible for data interoperability. The business layer helps yours organization to separate business logic from other logics and validations. The data access layer manages and stores the data into databases, files, warehouse (which is central repositories of integrated data from one or more data sources), and data lakes (which are storage repositories that have huge amounts of raw data). DXP design helps enterprises to design end-to-end holistic solutions that help business to maximize scalable and operational capabilities. We will look at these layers in detail in the next sections.

Presentation Layer

The presentation layer comprises user experience as well as UI. This layer helps you to understand the basic requirements as well as complexity of the application. User experience can make or break a brand's value. You need to consider the channels or touch points while designing the user experience journey, as shown in Figure 3-6.

Figure 3-6. *Touch points*

DXPs incorporate the design thinking approach to decide the touch points and user experience journey. This process consists of five stages: understand, define, design, prototype, and test.

- *Understand*: In this stage you gather the information to understand the problem you are solving using these touch points. Your assumptions become clear about the solution and you are able to get insight of the user's need regarding touch points and channels.

- *Define*: During this stage you put all the information gathered from the last stage in one place. You will analyze the observation and be able to define the problem clearly. You will have all the minute details required to design optimized workflow for the organization using the mentioned touch points and applications in Figure 3-6.

- *Design*: Having collecting all the information from the last two stages, you can start thinking about and identifying new solutions to the problem statement created during the design stage. In this stage you will be able to understand the technology stack and touch points needed to solve the problem, along with mockup user experience screens and UI components for touch points selected to provide a solution to the problem.

- *Prototype*: In the prototype stage, you create the inexpensive, scaled-down version with minimal features so that you can investigate the problem and its solution. You can consider this as a proof-of-concepts phase.

- *Test*: In this stage, you test the prototyped version of the solution. It is an iterative process; the result generated in this phase can redefine the problem and help redesign the solution.

Chapter 4 covers designing of the UI layer for mobile, tablet, and desktop in detail. Chatbot, VR, AR, Alexa, and voice assistance, as shown in Figure 3-6, enhance the user experience journey; these are the by-products (application) that are indirectly connected to the DXP's web application build for mobile, tablet, and desktop devices. IoT devices are considered smart devices that sense the user's input through sensors and pass the input to IoT boards; on the basis of input, further application would take action on the input and provide the appropriate output. We look into designing chatbot, VR, AR, Alexa, voice assistance, and IoT in detail in subsequent parts of this chapter.

While designing the presentation layer for a web application, you need to consider the different approaches. You look into the approaches like mobile first or desktop first. The DXP's design will follow the mobile first approach, as thinking mobile first will help you to design the content, which is really important for the application.

CHAPTER 3   DESIGN

Scripting Framework

Degree of modularity and maintainability should be considered while deciding the scripting frameworks to be used by your web application, for example, Angular or React Figure 3-7). Both are technology stack-adapted, component-based, modular programming approaches but both have advantages as well as limitations. For example, React has the advantage of virtual DOM (Document Object Model), which handles memory management efficiently, whereas Angular has the advantage of two-way binding.

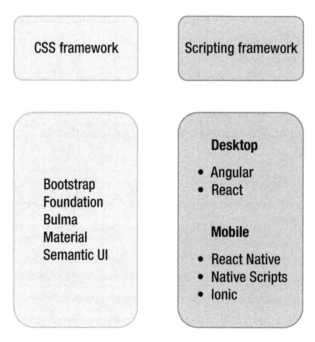

Figure 3-7.   CSS and scripting

React Native or Native Scripts are used for developing a hybrid application for mobile and tablet. A React or Angular technology stack along with CSS frameworks like Bootstrap, Foundation, etc. provide responsive UI that works on a mobile as well as a tablet's browsers. We look into these CSS and scripting technology stacks in Chapter 4 in details.

CHAPTER 3　DESIGN

UI Management

UI management provides the capability to organize and handle your UI scripts. It will help you to manage UI components and dependencies, and you can chose appropriate package manager, module bundler, task runner, and testing frameworks from Figure 3-8.

- *Package manager*: A JS package manager will help you to manage a package dependency that is installing, configuring, and removing dependency modules from a UI project.

- *Module bundler*: Using a bundler is the process of combining a group of modules (dependency) into a single unit in a specific order. You can use Webpack, Rollup, or Browserify as a module.

- *Task runner*: A task runner will help you to maintain their UI code by running different tasks like watching the files, minifying the files, lining JavaScript files, etc. You can use one of the task runners such as Grunt or Gulp, as shown in Figure 3-8.

- *Testing*: A JS testing framework will allow you to perform cross-browser testing; some frameworks provide both test environment and testing structure to your UI application. You can use the frameworks to generate, display, and monitor test results. The testing framework can be used in both environments: test-driven development (TDD), a process for when you write and run your tests, and behavior-driven development (BDD,) which lets you define application behavior in plain English text.

Steps to decide the UI design are as follows:

1. Select compatible CSS and scripting framework and technology stack (see Figure 3-7), appropriate package manager such as Yarn, NPM, or Bower, etc. (see Figure 3-8.

2. Select appropriate module bundler.

3. Select task runner, if needed.

4. Select testing framework such as Mocha, Jest, Jasmine, Cucumber, or Karma (see Figure 3-8) according to your chosen scripting technology stack such as Angular technology stack (ATS) or React technology stack (RTS).

CHAPTER 3 DESIGN

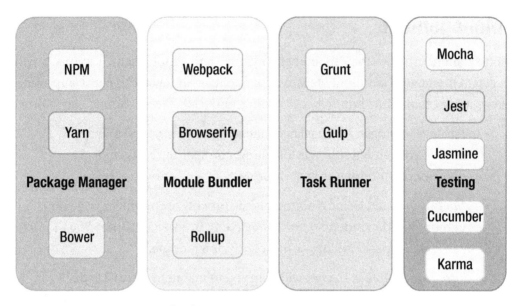

Figure 3-8. *User Interface (UI) Management*

UI Deployment

After building the DXP's UI application, you can deploy static content to the web server. The web server takes care of load balancing, proxy serving, web serving, security controls, and monitoring services.

Web and Http caches store the static items like HTML pages, images, etc. so that the application is able to deliver static content faster. You can use many options to deploy your static content, as stated in Figure 3-9.

CHAPTER 3 DESIGN

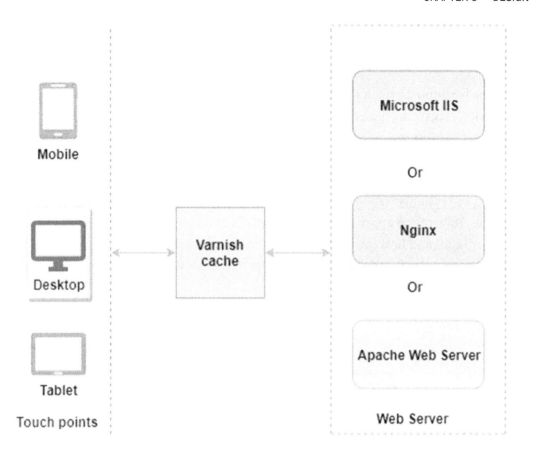

Figure 3-9. *HTTP accelerator, web cache, and web server*

Varnish cache along with the web server is a unique combination that will deliver content faster. You can choose any web server such as Microsoft IIS, Nginx, or Apache web server according to your needs and requirements. As shown in Figure 3-9, the client requests to Varnish and Varnish replies static content to the client or passes the request to the web server if the requested content is not cached in the Varnish cache. Varnish along with the web server is able to provide load balancing, proxy server, web serving, security controls, and monitoring services features.

Integration Layer

The DXP's integration design integrates data, applications, and people together through a common business process, for example, information portals, common business functions, service-oriented architecture (SOA), or distributed business processes. There are two types of integration: loosely coupled and highly coupled.

77

CHAPTER 3 DESIGN

Loosely Coupled Integration and Highly Coupled Integration

The loosely coupled principle reduces speculation between components and applications regarding their exchange of information in the form of messages (Figure 3-11). Messages are sent in a particular format such as JSON or XML. It is asynchronized, whereas tightly coupled solutions are synchronized in nature and create interruptions when changes are required (Figure 3-10). Before an exchange of data message, the system establishes connection using Connect and Ack (acknowledge) messages. Integration approaches have evolved significantly from RPC (remote procedure call) and RMI (c), supported with many platforms and frameworks like CORBA, Java RMI, RPC-Web services, SOAP services, monolithic REST services, and microservices.

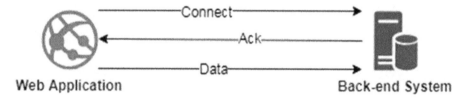

Figure 3-10. *Highly coupled*

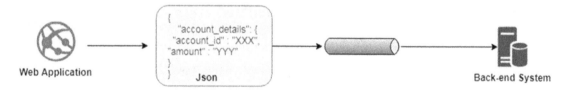

Figure 3-11. *Loosely coupled*

Integration helps you to integrate multiple platform using integration patterns and solutions. Key integration patterns are as follows:

- *Channel patterns*: These patterns provide the way to transport the message across a channel. Patterns such as point-to-point channels, publish-subscribe channels, message bus, etc. are shown as message channel in Figure 3-12.

CHAPTER 3 DESIGN

- *Routing patterns*: These patterns provide the way to route a message from sender to receiver, shown as Message Routing in Figure 3-12. These patterns consume the message from one channel and send it to another channel without modification on the basis of a set of conditions.

- *Transformation patterns*: These patterns change the content of a message, for example, XML to JSON conversion (also known as message translation).

- *Endpoint patterns*: It is a messaging system so that a client can consume or produced messages. It defines endpoints, which are consumed by other applications. It has Message Endpoints, Message Gateways, and Message Dispatcher, as shown in Figure 3-12.

- *Management Patterns*: These patterns help to deal with errors, performance analysis, and logging in the application.

- *Message construct*: It has the message encapsulated with data, along with message events. It is responsible for construction of the message and holding the return address.

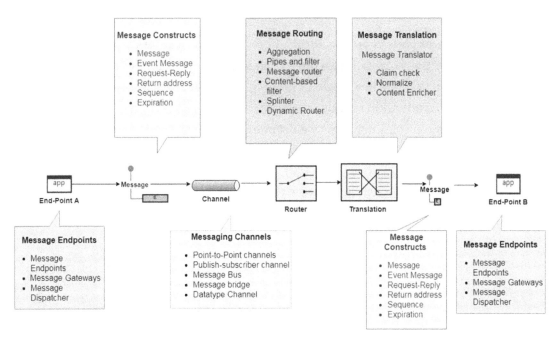

Figure 3-12. *Integration Components and Patterns*

CHAPTER 3 DESIGN

The patterns mentioned will help to build services and solve integration problems from End-point A (application A) to End-point B (application B). These patterns encapsulate the design knowledge; hence, irrespective of any integration technology, it will help to solve the integration problem.

We look at the following message patterns and message components in detail in Chapter 5.

- Message
- Message channel
- Pipes and filters
- Message router
- Message translator
- Message endpoint

Following are different integration architectures: for example, information portals, common business functions, and distributed business processes architectures, which help you to solve common problems of integrations such as aggregator pattern architectures, B2B (business to business) architectures, and service bus architectures (SOA).

- *Aggregator pattern architectures*: There are many business processes where you should to access more than one system to answer a question to perform a single business function in an organization. Hence, information portals aggregate the information from multiple sources to display a holistic view. For example, you need to get data from business process 1, business process 2, and business process 3, aggregate the data from all processes, and display it on the dashboard, as shown in Figure 3-13.

CHAPTER 3 DESIGN

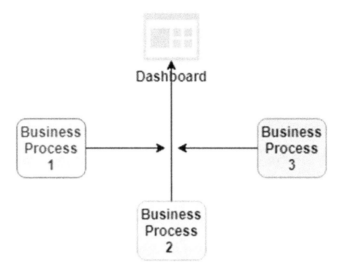

Figure 3-13. *Aggregator pattern*

- *Business to business (B2B) architectures*: There can be a scenario where you need to integrate business functions or processes available from third-party suppliers or business partners; for example, a bank provides billing and recharge functionality, hence the bank needs to integrate utility services from third-party suppliers. In this architecture we are integrating two system or business processes outside the organization. These systems are directly communicating with each other.

- *Service bus architectures (SOA)*: Shared business processes and functions also referred to as services. Once an organization collects a set of services, service bus architectures provide tools that make calling an external service almost as simple as using conventional methods. In this architecture all the services are communicating with the DXP's application using a common bus, as shown in Figure 3-14. Services from external system 1, external system 2, and external system 3 communicate with the bus, and these services are exposed to the DXP application via common bus.

CHAPTER 3 DESIGN

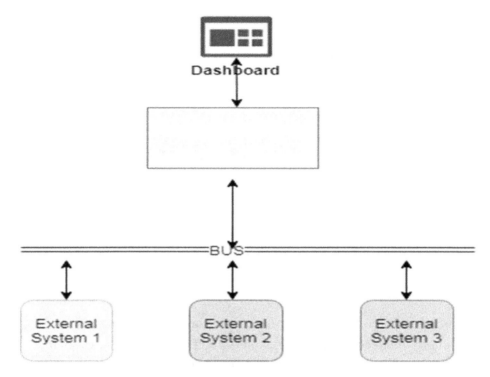

Figure 3-14. *Service bus architectures (SOA)*

Chapter 5 covers designing of the integration layer in detail. Integration deals with data interoperability between different applications within or outside an organization, using different protocols and data formats. We delve into micro services, REST services, ESB, and API gateways in Chapter 5.

Figure 3-15. *Integration*

The DXP's architecture supports microservices and monolithic services architecture because it makes integration structure more flexible, as it structures the application into multiple modular services. ESB and API gateway help you to scale your services build using microservices and monolithic architecture as shown in Figure 3-15. Microservices are lightweight services running as a separate process; each service inculcates a separate business capability in it. The advantages of microservices over monolithic applications are as follows:

- Services communicate using REST.
- Services are loosely coupled.
- Scaling is easier.
- You can achieve isolation of services: if one service fails, another would continue.

You can build these microservices using containerization and orchestrate using Kubernetes. The advantages of microservices build using Kubernetes and containers are as follows:

- Number of services can be deployed and delivered quickly.
- Services built on Kubernetes can be deployed across different environments.

Test environments like SIT, UAT, or preproduction can be cheaply and quickly created with a Kubernetes cluster.

Steps to design the integration layer are as follows:

1. Select the appropriate architecture type such as microservices or monolithic services architecture.

2. Select the appropriate messaging pattern for the business process to be integrated.

3. Select the appropriate framework that satisfies the business needs; for example:

    - Apache camel can be used as mini-ESB for large scale applications.
    - Apache CXF can be used as service framework for medium scale applications.
    - Spring Boot along with Apache camel can be used for microservices.

CHAPTER 3 DESIGN

4. Get mutual consensus on authentication details, data formats, services endpoints, and services protocols from all systems participating in integration; for example:

- Get the data formats details, like services integrated will have XML or JSON.

- Get the service endpoints and details like service URLs, and port number and its invocation type (verb) or method; for example, POST, GET, PUT, DELETE.

- Get the service protocols details such as RESTful or RESTless.

- Get the authentication details and encryption-decryption algorithm used to secure data while communicating with two systems; for example, authentication token passed along with these services.

Business Layer

A DXP provides a separate layer for business logic so that integration and application logic won't be able to hamper business logic. As shown in Figure 3-1, the business layer has a business controller, business validation, and data transfer object (DTO)

Figure 3-16. *Business layer*

The business controller is responsible for receiving and replying to requests. It is responsible for the following:

- *Redirection*: You can redirect the control of an application to business services.

84

- *Business logging*: You can log the request and response in the business controller of the application.

- *Authorization*: You can integrate authorization logic, which will check the user's authorization to access the data from a particular business controller.

Business validation or services are responsible for filtering and processing a DTO received from the data access layer on the bases of business rules and logic. After processing, objects are sent back to a particular business controller, which has initiated the request for business services. Business services are responsible for the following:

- *Model binding*: Data received from Data access layer will be mapped to business services.

- *Business rules validation*: Business rules and checks like null pointer exception and blank check, etc. are validated in business services.

The data transfer object interacts with the data layer (database). Records stored in a database are mapped to entity objects, and entity objects are requested to appropriate tables in the data repository. Entity objects are then sent back to business services where objects are processed, then these processed objects are sent back to the business controller, as shown in Figure 3-17. The business controller redirect to appropriate business services, and business services access the data received from the DTO and data access object (DAO) of the data access layer.

CHAPTER 3 DESIGN

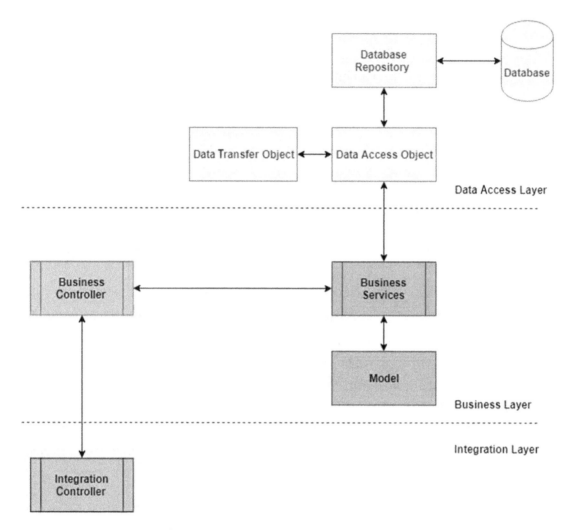

Figure 3-17. Business layer to data access layer

Data Layer

The data access layer is responsible for simplified access to data stored in databases, as shown in Figure 3-17.

This layer contains a DTO, which is a mapping to the table; every column in the table is a member of the DTO. The DAO (data access object) helps you to create, delete, modify, or search for an entity using a simple object.

The DAO design pattern is used to implement the data persistence layer. It is based on abstraction and encapsulation design, as it protects other parts of the application from any change in the data layer, for example, change of database from MySQL

CHAPTER 3 DESIGN

to PostgreSql or from database to file system. As an example, a DXP's application authenticates a user by utilizing the database but it is later decided to go for SSO (single sign-on) using LDAP and SAML. It would be safe if the user were using DAO to access data from the database, as the user only needs to make changes on the data access layer. Java-based projects use JPA and hibernate framework to access data from the databases. You can access data from files, data lakes, NoSQL databases, SQL databases, etc., as shown in Figure 3-18.

Figure 3-18. Data

Middleware Layer

Middleware is an infrastructure that helps to deploy and manage complex business applications. Components of a DXP's applications may be developed using various programming languages, protocols, and frameworks. Middleware provides the services and helps in implementing NFRs (Figure 3-19) such as transaction concurrency transaction monitoring, security authentication and authorization, transaction logging, transaction auditing, and distributed processing.

Figure 3-19. Middleware components

87

CHAPTER 3 DESIGN

Middleware represents a collection of interconnected components that are distributed across different locations and provides features like reliability, scalability, and maintainability. Therefore, middleware makes application access easier. It provides load balancing servers, web servers, and application servers. CMS facilitates middleware infrastructure and supports application development and delivery.

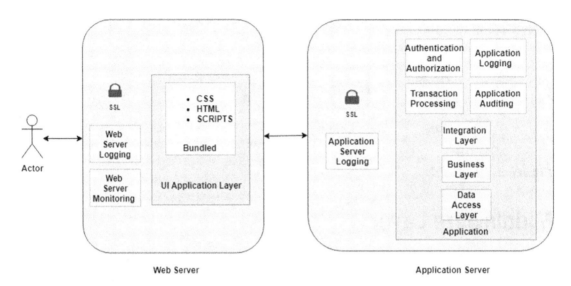

Figure 3-20. *Middleware layer*

You should look into the various components of the middleware layer carefully while developing and deploying your DXP's application, such as application monitoring, server monitoring, application logging, server logging, and auditing. As shown in Figure 3-20, the UI (front-end) applications will be deployed on a web server and the backend applications will be deployed on the application server. Backend applications have different layers and components, such as authentication and authorization, logging, and auditing.

- *Application monitoring* ensures that application processes perform in an appropriate manner.

- *Server monitoring* ensures health and availability of the server and OS; that includes bandwidth, CPU utilization, memory utilization, and disk utilization.

CHAPTER 3   DESIGN

- *Application logging* ensures that logging errors, information events, and warning are appended into log files. Logging helps to check the issues reported by any users.

- *Server logging* retains the logging errors and information events and warnings generated by the server. The server monitors its own list of activities.

- *Auditing* ensures that all the activities done by a user on an application are logged in the auditing file or database.

- *Transaction processing* initiates and interacts with the integration layer and business layer. It ensures the reliability and consistency of any kind of transaction that takes place in the system; in case of failure, it will roll back the failed transaction.

Social and Collaboration Design

You can integrate collaboration tools like Gerrit, Jira, Rocket Chat, Slack, and Yammer with the DXP application. These tools have RESTful APIs available, which can be integrated easily with the DXP application.

You can integrate with social platforms, for example, Twitter and Facebook. These tools have authentication API available, so that you can authenticate the application using these APIs. You can use Facebook and Twitter APIs to integrate tweets and posts in your own application, as shown in Figure 3-21. The DXP UI layer gets the authenticating details from the user, and these authentication details are passed to a third-party social platform such as Facebook or Twitter. These platforms provide an access token to further interact with these platforms and get the users data to your DXP application.

89

CHAPTER 3 DESIGN

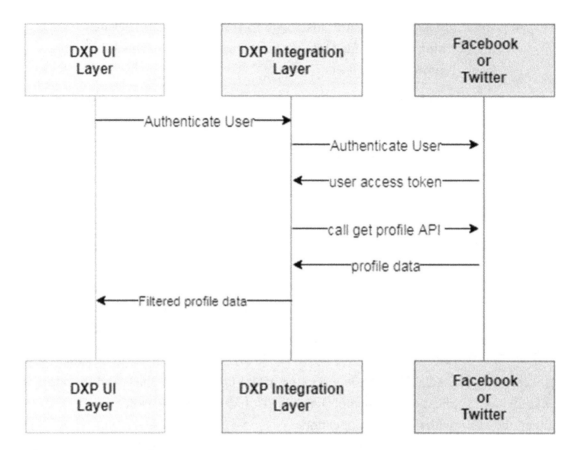

Figure 3-21. *Social integration*

Collaboration is all about conversations between people to get to a goal. It is about cross-questioning, collecting answers, and getting feedbacks. It is all about social interaction, one of the ways that work gets done. The traditional way of business collaboration was e-mail with attachments, but collaboration was slow, difficult, and inefficient. Therefore, to solve the collaboration problem, social software solutions, based on microblogging, social networking, and wikis were integrated with DXP.

CHAPTER 3   DESIGN

Figure 3-22. *Social and collaboration requirements*

Social software applications like Instagram, Facebook, and Twitter have fast gained adoption, considering there is a strong social incentive to use them. When people are utilizing an application to reach out to you, then you should build social networks for that; you need to choose a solution that incorporates microblogging, social networking, dynamic profiles, and automated activity feeds. You need to decide on many factors on which social software solutions will be integrated with a DXP. You can integrate wiki for information sharing, Facebook authentication, and feed APIs to integrate user social interaction with the DXP; that will further enhance the user experience. You can implement a rule-based chatbot for automating frequently asked queries by the user.

- *Live chat*: Live chat is one of the customer servicing tools. It provides chatting capabilities in real time via a chat window placed in the website or mobile application.

- *Chatbot*: You can integrate a chat engine with a DXP application. Chatbot engines are of two types: rule-based engine and ML prediction-based engine. It depends upon your requirements and workflow which engine you want to integrate with DXP, as shown in Table 3-2.

  Table 3-2. ChatBot

Engine	Usage
Facebook Bot	Facebook messenger interaction
Slack Bot	Automate developer team interaction
Chat fuel	Bot for marketing, sales, and support
Dialog flow	AI based bot

- *Wiki*: You can integrate an open-source enterprise wiki—for example, TWiki—with a DXP application, which will enhance collaboration of teamwork together seamlessly and productively.

- *Blog*: You can integrate a blogs framework, for example, WordPress with a DXP application where people can share knowledge between teams.

- *Calendar*: You can integrate a Google calendar API with a DXP, so that you can plan your work and interact with the Google calendar.

- *Forums*: A forum is a type of message board, divided into topic folders, where you can publish posts and reply to posts from other team members, for example, vanilla forums.

- *KM portal*: Knowledge management portals are considered to be virtual workplaces that promote knowledge sharing among different categories of end users and provide access to stored structured data and organize unstructured data, for example, Plumtree and Woolmani. You can integrate this portal with a DXP to manage an organization's knowledge.

- *External integration with FB, LinkedIn, and Twitter feeds*: You can integrate social media API to a DXP application where you can integrate feeds and SSO authentication.

IoT Integration Design

IoT is another fast-growing technology, which can assist you to build and implement a data gathering network to improve systems and establish a new channel for interaction. DXP is a platform that is capable of quickly adapting emerging technology to address major challenges. DXP is leading organizations toward innovation, which helps businesses with intelligence and advice.

IoT systems use sensors to provide operational insights from the data. Intelligence is added by integrating the analytics and ML into IoT devices. The IoT is effectively connecting the digital world to the physical world. It helps to communicate and integrate information systems and the fields' data, and it is possible to use this data in real time. The IoT is mainly divided into three layers as shown in Figure 3-23: physical sensing layer, IoT Integration layer (also called as middle layer), and IoT application layer.

- *Physical sensing layer*: Physical sensing includes sensors, for example, temperature, proximity, pressure, etc.; and development boards also called prototyping boards to acquire data from the environment, like Arduino, Raspberry, Edison etc. Along with mobile phones and their sensors like microphone to get voice commands, proximity, accelerometer, and other sensors can be grouped as passive or active (passive sensors don't require external power sources, whereas active sensors require external power to sense the external environment). Analog sensors produce a continues signal, whereas digital sensors produce a discrete signal.

- *IoT integration layer*: The IoT integration layer integrates the data collected from a device to a NoSQL database along with a distributed computing platform like Apache Spark and Hadoop. Data acquired from sensors can be stored in the cloud and can be used later to create dashboards. The IoT integration layer is also called the middleware layer. This layer ensures security, quality of service (QOS), and provides IoT gateways as well. You can implement this layer in your DXP application by integrating IoT frameworks such as Iotivity, AllJoyn, Eclipse-Kura, etc.

CHAPTER 3 DESIGN

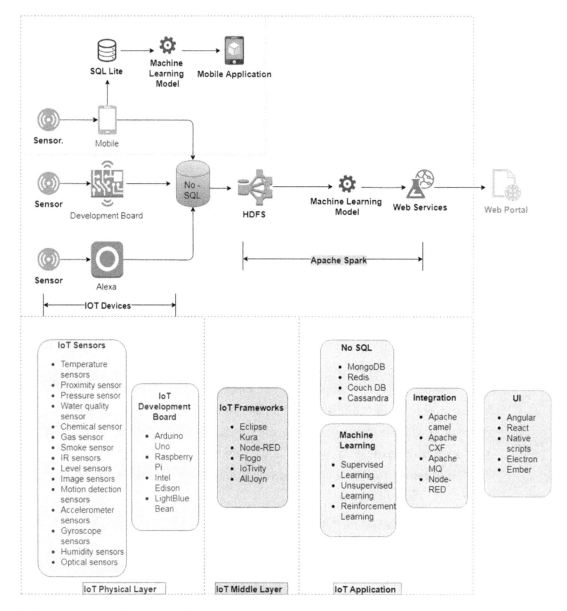

Figure 3-23. IoT integration

- *IoT application layer*: The IoT application layer collates the data and applies ML capability to the data taken from the devices. Insights, predictions along with appropriate data, and web services are exposed to the UI dashboards and analytics application. Frameworks like Apache Spark and Apache Kafka help to manage IoT data in real time; they provide data pipelines and steaming mode to get insights from an IoT application for the organization. In the layer, you can use TensorFlow, Apache OpenNLP, Apache Tika, and other ML and NLP libraries for sentiment analysis, image analysis, document analysis, time-services analysis, and other processing. IoT data can help recognize trends common in a machine that help us to understand the breakdown of applications.

At each layer in the design, you can run various ML libraries as needed. In Figure 3-23, IoT Integration has four open-source software components designed to collect the data from external environment (i.e., sensors and boards); data stores (i.e., NoSQL databases); management (i.e., integration of IoT framework and platform), which provides the communication to collect data from a group of sensors from multiple locations, visualization (UI applications), and manipulation (ML) of time series IoT data in an easy and scalable manner. IoT frameworks like ARIoT, AllJoyn, Iotivity, etc. have implemented SOA, which is beneficial in IoT integration. These frameworks provide data interoperable capabilities, the API layer. Physically sensed activities generate the events, and these events are sent via the communication protocol such as MQTT to the service (middleware) layer. These events are pushed via the event bus to a REST API and then the update is reflected on web portals and mobiles devices.

IoT Case Study

Integration of technology such as AI/ML and big data along with blockchain will prove to be extremely useful for IoT use cases in the near future. The following use cases will help you to understand the IoT applications and implementation so you are able to implement IoT-based applications in your organization.

- *Asset tracking*: Logistics organization already has tracking assets using IoT. Organizations can have real-time access to the appropriate location of the assets. You can attach tagging sensors with the assets that will improve efficiency, as it will resolve the problem of locating the assets. IoT-enabled things assist people to improve productivity.

- *Smart cities and real-time streaming data*: Smart cities are based on connected technology that makes cities more progressive and have data to get insight that can help in improving safety, economy, and quality of life. Real-time data streaming through streaming engines, data pipelines, and IoT will lead us toward smart cities. A citywide information network could be linked to sensors and a digital experience platform that enables the city to provide automated street lighting, waste management, digital bus routes, and smart parking.

- *Digital wallets*: The IoT can extend the capabilities to automate payments through devices such as digital wallets, which can be attached to each device. For example, digital wallets attached to cars can pay for fuel charges, road taxes, etc.

- *IoT smart payment contract*: Smart contracts are computer programs that verify and enforce the negotiation of a contract. Data captured from IoT devices can execute the smart contract and the system would deduct the payment by coordinating with the bank.

- *Banking through wearable*: The ecosystem of IoT is growing day by day. Many banks have started providing services through wearables like the Apple watch, etc. Applications could be built for already existing wearable devices for contactless digital payments. Integration of IoT devices with a digital experience platform enhances usability.

Blockchain Design

Blockchain could be used to record transactions or events, which would be replicated exactly across all the nodes in a network. Every node would have a copy of records that cannot be edited or deleted. We will look into Blockchain concepts, smart contracts, the Blockchain platform and its design components, along with a use case study in this section. Let's look into Blockchain concepts using a book and library analogy.

What is Blockchain?

Here is an attempt to understand blockchain with the analogy of book and library. It is a known fact that an accounting book is called a ledger, every page in the ledger is connected to other pages in sequential order, and these pages contain transactions. Pages can be considered as blocks and the book can be considered as a blockchain.

CHAPTER 3 DESIGN

What Is a Distributed Ledger?

A replicated and consistent version of a ledger is distributed across libraries. Libraries can check and validate the authenticity and consistency of a book by observing, comparing, and taking consensus from other libraries. If a ledger is distributed across libraries, this can be considered a prefect analogy of a distributed ledger. A distributed ledger wouldn't have central trust authority.

Libraries are considered as nodes. Each node has a consistent version of a ledger. Transactions are added in blocks and blocks are added in the ledger, with the consensus of nodes participating in blockchain network (Figure 3-24).

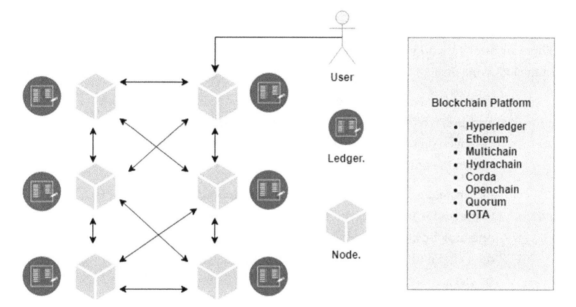

Figure 3-24. Blockchain

Smart Contract

Smart contracts are self-executing computer programs with an agreement between the participants on assets, and this contract exists across a distributed and decentralized blockchain network. For example, in a banking use-case, an account can be considered as an asset, whereas banks and account holders are the participants. Participants are the users of the blockchain network. Participants can write a smart contract with predefined agreements on assets. On the basis of transactions done by participants in the network,

97

the smart contract is executed; all the events along with the transaction are recorded in the network and stored in a blockchain. These transactions are appended in the blockchain, hence immutable in nature.

Blockchain Platforms

A DXP provides the capabilities to deploy and integrate the different components of the DXP application with the blockchain platforms, that is, Hyperledger, Etherum, Multichain, Hydrachain, Corda, Openchain, Quorum, and IOTA, as shown in Figure 3-24.

DXP and Blockchain Network

Three kinds of networks are involved while integrating blockchain with a DXP, as shown in the following design.

- *Enterprise network*: Existing applications reside in your enterprise network, which contains the organization's data that may be deployed on stand-alone infrastructure or cloud infrastructure, as shown in Figure 3-25.

- *Public network*: An organization's data is exposed using REST APIs to the public network, where users (analyst, administrator, auditor, operator, business user) of the application can utilize the authorized data after authentication in the form of an analytic chart, monitoring services, API management, and Blockchain explorer (see Figure 3-25.

- *Blockchain network*: The blockchain network consists of blockchain components, that is, smart contract, ledger and transaction, e-certificate, membership services, public–private key infrastructure, and interoperation, that contain events and communication protocols (see Figure 3-25).

Blockchain is of two types, that is, public blockchain and private blockchain. In case of a public blockchain, anyone can be allowed to participate in the network, can execute the network, and maintain the ledger; in a private blockchain, identity services maintain the roles and responsibilities, so that only members of the blockchain can execute the network and maintain the ledger. In addition, a DXP supports integration of an open-source private business blockchain network.

CHAPTER 3   DESIGN

Blockchain Components

A blockchain platform has multiple components to establish and manage a blockchain network and deploy smart contracts on it. We will look into these components as follows, such as consensus layer, smart contact layer, network communication layer, data store layer, crypto layer, and services such as identify management and API management, as shown in Figure 3-25.

- The *Consensus layer* is responsible for collecting valid transactions in a block and appending a new block in the blockchain network after taking consensus from the node participating in the network. You can use different kinds of consensus algorithms (POW, POS, DPOS, POA, PBFT, BFT) according to the chosen blockchain platform.

- The *Smart contract layer* is responsible for processing transaction requests and determining if transactions are valid or invalid by executing business logic. The smart contract is developed and deployed by the blockchain developer in the blockchain network.

- The *Communication layer* is responsible for peer-to-peer message transport between the nodes that are participating in the network. It supports the interoperation among different blockchain instances.

- The *Data store layer* allows different data stores to be used by other components in modules.

- The *Crypto layer* contains different crypto algorithms to be swapped out without affecting other modules and layers.

- The *Identify services* contains E-certificates and public key infrastructure (PKI), which establishes the trust during setup of blockchain instances. It interacts with member services for enrollment and registration of identities or system entities during network operation. It also provides authentication and authorization to access the events and transactions in the blockchain network.

- *Data services and API managements* enable clients and other DXP applications to interact with the blockchain application and network.

CHAPTER 3   DESIGN

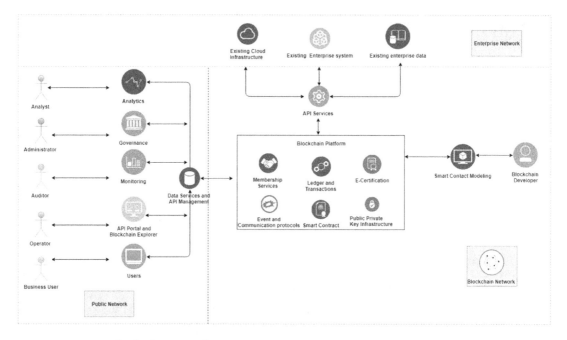

Figure 3-25. Blockchain architecture

Blockchain Case Study

There are many use cases that can be solved using blockchain or distributed ledger technology, which can ensure immutability and transparency of records in the ledger and provide a common platform for users across industries and organizations. Use cases such as KYC, EHR, digital identity, POE (proof of existence), claim management, etc. are explained in brief in following section.

- *Know your customer (KYC)*: The blockchain-based KYC process using digital identity helps banks to know their customers. A digital signature is created and stored in a blockchain-based system; data is linked with the customer's digital signature, which is decrypted with the customer's private key.

- *Electronic healthcare records (EHR)*: Blockchain is used to provide patients an ecosystem to control and store their health records in a blockchain network. Patients could give or revoke permission to share their medical record with a medical institution and doctors, although every medical institution would be appending medical records and tests in the user's blockchain in a distributed ledger.

- *Digital identity*: A blockchain-based data encryption digital identity management platform could defend against identity theft. One could choose which data to share with whom across different channels.

- *Letter of credit*: You can use blockchain and deploy a letter of credit as a smart contract between the bankers or investors and the supplier to guarantee payment. If the products and services are delivered according to the buyer with all specified conditions in the smart contract, the contract gets executed based on the documents submitted by the various parties verifying that the letter of credit conditions meet specified shipment deadlines and conditions. This can be automated through program logic in the smart contract to indicate and check compliance or noncompliance.

- *Proof of existence*: PoE is also called an authenticity of a file or record on a blockchain. The publisher or creator of the document uploads the file along with its hash to the blockchain network; the verifier can check its authenticity by uploading the document. The blockchain application calculates the hash of the document uploaded by the verifier and matches the hash calculated with the hash of the document available in the blockchain network.

- *Claim management*: You can easily audit and maintain transparency while making claims. A smart contract can be modeled according to claim process and deployed on the blockchain network. Claim smart contracts automatically and securely complete all the steps involved from automating coverage verification to claims validation.

- *Loyalty and rewards*: A blockchain-based loyalty and rewards platform provides and maintains transparency among stake holders. Multiple reward programs and cards are merged in one platform and then rewards can be used anywhere without any restrictions.

CHAPTER 3 DESIGN

Big Data and NoSQL Design

Big data analytics can provide and uncover the patterns hidden in your organization data. You can integrate actionable insights with DXP and create efficient data streams that can learn, predict, and take action by connecting DXP with multiple data sources, and then apply ML algorithms for better understanding of their own customer.

Big Data and NoSQL Integration

Big data solutions are built using open-source projects like Apache Spark, Hadoop, and Kafka, to name a few, which help you to collect data from multiple data sources and provide distributed processing, and usage of ML algorithms and data visualization methods help you to analyze big data that helps management as well, as shown in Figure 3-26. We will look into big data components such as ETL, ML models for efficient steaming of predictive data models, search and query web services, and usage of NoSQL databases in this section.

- *Extract, transform, and load (ETL)*:
  - You can load data from multiple data source using open-source big data streaming engines such as Apache Spark. It can access multiple data sources including the Hadoop Distributed File System (HDFS), NoSQL database, and SQL-databases.
  - Collection of elements of your dataset that will be stored in memory or disk across a cluster of machines
  - A data frame is created to help process large data sets easily. Spark's dataset and data frame provide an API that allows developers to easily express transformations on domain objects.
- *Train and test predictive data model*:
  - You can use different kinds of ML algorithms (supervised learning, unsupervised learning, or reinforcement learning) depending upon the nature of problem.

- You can use Spark's ML.lib or other ML libraries such as Tensorflow, PyTorch, etc. that allows data scientists to focus on their data problems and models instead of solving the complexities of distributed data, such as infrastructure and configurations, etc.

- Machine learning algorithms involve a sequence of tasks, including preprocessing, feature extraction, and model fitting, in identifying outliers. In the case of Apache Spark, ML Pipeline is a high-level API for ML provided by Spark that provides a sequence of stages handled with distributed processing capabilities.

- *Data streams and processing*:

  - Data stream processing helps data engineers and data scientist to process real-time data from sources including stream engines such as Apache Kafka, Rabbit MQ, Redis Simple Message Queue (RSMQ) and Flume.

- *Search and query web services*:

  - Processed data can be pushed out to file systems, databases, and live dashboards using web services.

  - Web services are exposed to the UI dashboard, as shown in Figure 3-26. You can trigger a query using a Web API. These WEB APIs further interact with an ML-based trained model; the model loads and processes the real-time data and returns prediction results back to databases and UI dashboards.

CHAPTER 3   DESIGN

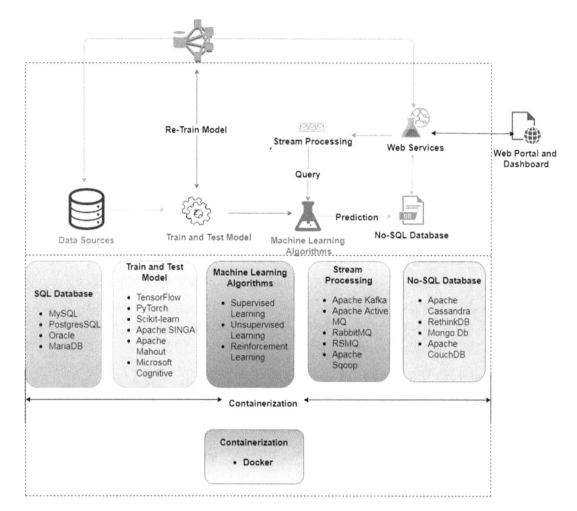

Figure 3-26. *Big data and NoSQL*

- *NoSQL database*:

    - A NoSQL database is recommended with an analytics application, which receives huge amount of data in real time and needs to update the dashboard with the latest data. Searching and querying a NoSQL database is more efficient than an SQL database. Table 3-3 lists the advantages of using NoSQL in big data solutions.

Table 3-3. *SQL vs. NoSQL*

SQL	NoSQL
Relational database	No relational database
Fixed schema	Dynamic schema
It is a table-based database.	It can be a collection of key-value pairs, documents, and graph databases.
MySQL, SQL Server; etc.	Mongo db, Couch db; etc.
It has defined SQL language to define and manipulate the data.	It has unstructured query language used to query the data from collection of documents.
Vertically scalable	Horizontally scalable
It is used with transaction-based systems and solutions.	It is used with mobile applications, real-time analytics, and content management systems, etc.

- *Containerization*:
  - The application is built using a variety of frameworks, libraries, tools, and technology, which is encapsulated in a single container along with its environment. The application container is deployed on multiple virtual machines (VMs), cloud infrastructure, or on a standalone machine.

Big Data and NoSQL Case Study

Let's look into big data use cases that can be achieved by the aforementioned design.

- *IoT model-based algorithm*: An organization can use an IoT model along with ML algorithms to learn from historic events and make smart decisions. This helps financial institutions to make smarter investments. You can make innovative use of big data and IoT. For example, a bank can use behavior analysis to analyze customers' visits and money transactions from different bank branches; integrating these analyses into the business model helps to create customer-centric deals, personalized offers, etc.

- *Employee engagement*: You can apply big data analysis to track performance of employees. Analytics could ensure and understand employee productivity.

AI Automation Design

AI automation is comprised of traditional automation and robotics process automation (RPA); RPA is an emerging technology. Traditional automation is the automation of any type of repetitive task. It is usually found in a workflow-based application, whereas RPA allows organizations to automate tasks like the way human beings interact across the system and application. The main goal of RPA is to replace repetitive tasks performed by humans with a virtual workforce. AI has been categorized and grouped, such as RPA, speech reorganization, NLP, deep learning, ML, and chatbot (virtual bot). You need to determine the automation goals, followed by building the AI model.

Determine Automation Goals

You need to decide an automation strategy that clearly sets out how and where you apply automation. After determining the automation strategy, you can map it with predefined automation systems, which will further help to decide the framework and algorithms to be integrated with DXP, for example, NLP-based solutions or data prediction-based solutions.

Steps to Build AI Automation Model

The approach to build an AI model is as follows, where you create training and test data sets and apply ML algorithms or neural network algorithms to build an AI model. You have five steps: data preprocessing, build the model, train the model, test the model, and improve or tune the model on the basis of expected results from the model, as shown in Figure 3-27.

- *Data preprocessing*: Data preprocessing is the initial step toward building an AI-based model. Datasets are loaded, cleaned, and processed according to the problem statement.
- *Build the model*: The model is built using ML, deep learning algorithms, and neural networks concepts, which further use ML libraries, for example, tensorflow, pytorch, etc.

- *Train the model*: The training data set is identified and used to prepare a model.

- *Test the model*: This is a new dataset, different than the training set; you gather predictions from the trained model with the inputs from the test dataset and compare them with the withheld output values of the test set.

- *Improve and tune the model*: You can adjust various parameters and tune the weights to improve the model built.

Figure 3-27. *AI automation model*

Chatbot Case Study

Most common AI solutions are built using Tensorflow, Python, and Spark. AI strategy helps to solve defined business problems, with a defined data set to solve the problems. Chatbots are programs built with NLP, which is supposed to solve domain-specific problems and query request by simulating human conversation. There are three components: presentation layer; bot layer, which has the bot framework or engine; and transaction and data processing system, which interact with the DXP to integrate the existing system with the chatbot engine, as shown in Figure 3-28.

- *Presentation layer*: The chat interface can be a custom UI (e.g., Angular, React) and native mobile application. You can interact with this layer using text, voice, and visual. Inputs are sent to the backend layer using web socket communication and REST APIs.

- *Bot engine*: This is an open-source bot framework used to create the bot model for specific use cases and domains to understand the intent of the user. The bot framework has capabilities to process, understand, and generate language that is NLP, NLU, and NLG as follows.

    - *NLP (natural language processing)*: This component understands the text or voice and understands the intent of the user.

CHAPTER 3 DESIGN

- *NLU (natural language understanding)*: This component helps your application to understand the intent and take action on it. After understating the intent, you can call APIs to interact with the external system and get or put information in other systems.

- *NLG (natural language generation)*: After understating intent and getting external information from other systems and databases, you need to generate the resultant message and send it back to the client.

- You can use open-source bot frameworks such as botpress, WIT.ai, etc. or use chatbot platform services like chatfuel, dialogflow, etc. to build a chatbot, or you can your create a custom framework using ML and NLP libraries with the help of a recurrent neural network and bag-of-words model.

- *Integration with legacy system*: If you are building a chatbot for a business, then most likely you are working with CRM, ERP (enterprise resource planning), and core banking, etc. You need to integrate the chatbot with an external system using REST APIs (external).

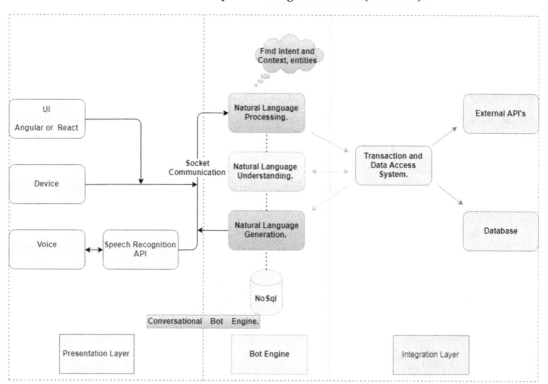

Figure 3-28. *Chatbot Integration with DXP*

Enterprise Search Engine

An enterprise search engine is used to search content from multiple sources (databases, files, and intranet) within the organization. Components of a search engine are as follows:

- *Processing*: Diversified data loaded from different sources will have different formats. This component processes the incoming documents to plain text and normalizes to improve precision and recall value, which includes stemming, that is, reducing words to their stem such as "texting" becomes "text"; lemmatization, which is the process to reduce the word into its base dictionary word such as "studies" becomes "study"; part of speech tagging, etc. An analyzer is used to analyze data and give back meaningful terms or words.

- *Indexing*: Processed text is stored in an index, which is used for quick lookup and will be handled by indexer. The dictionary contains an index of all unique words as well as information about their ranking.

- *Query processing*: A user from the web application executes the query. The query is broken into terms and operators using a query parser and analyzer.

- *Matching*: The processed query is compared with the stored indexes in the dictionary.

Now, we look at an enterprise search engine in detail, based on Apache Lucene and its technology stacks: Elastic Stack and Solr Stack. Apache Lucene is able to achieve fast search responses because it searches indexes instead of searching whole text.

- *Elastic Stack*: The Elastic technology stack has multiple components available to build solutions on top of it, such as Beats, Logstash, Elasticsearch, and Kibana, as shown in Figure 3-29, are as follows:

  - *Beats* collects the data and parses it and pushes it to Elasticsearch for log analysis. Log analysis can be achieved using Logstash and Kibana.

  - *Logstash* can connect to a variety of sources such as Web API, social services, IoT sensors, and databases and data streams like Kafka or Redis, which collect the data and pipeline to Elasticsearch.

Chapter 3 Design

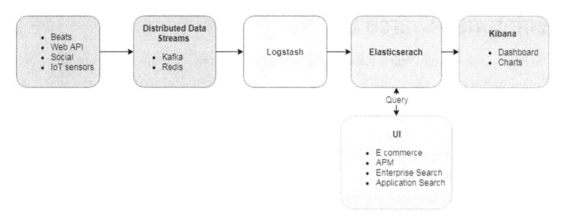

Figure 3-29. *Elastic Stack*

- *Elasticsearch* is a search server based on Apache Lucene. It provides distributed full-text search engine capabilities with RESTful web services. It stores and indexes the data.

- *Kibana* visually explores the data by querying Elasticsearch, or you can use their custom UI to fulfill the search requirements on the basis of their domain and fields. It is an analytics and visualization platform, which provides dashboards and charts for visualizing the data as per the search query.

Elastic stack can be deployed on Elastic Cloud or can be deployed as a standalone cluster. It can be used in e-commerce applications for filtering the data by end user, such as filtering by brand name and other features of the product. It can also be used for application performance monitoring (APM); log data from the application server can be loaded to Elasticsearch using Beats, and key performance indicators (KPIs) analysis will be displayed on the dashboard using Kibana.

- *SolrStack*: The Solr technology stack has multiple components available to build solutions on top of it, such as Logstash, Apache Solr, and Banana as shown in Figure 3-30:

  - *Apache Solr* collects the data from different data sources through the connectors; the data source can be data streams, files, application databases and documents, etc. Solr provides the parameters required (data) for visualization to Banana.

- *Banana* is a fork project of Kibana, which works with Apache Solr and provides visualization and exploration capability. It provides rich and flexible UI, which enables the user to develop an end-to-end search application. It also has a tabular display to drill down to the documents in a result set.

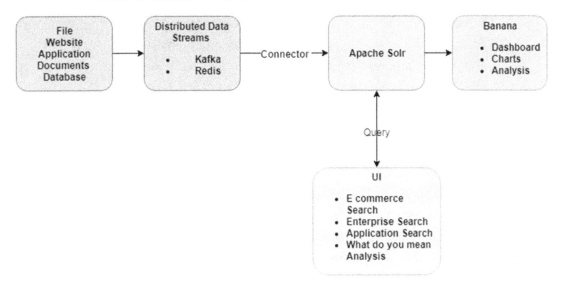

Figure 3-30. Solr Stack

You can create custom UI components according to their domain needs. Solr Stack can be deployed as standalone or in cluster mode.

Augmented – Virtual Reality Integration

Augmented reality comprises two layers: presentation layer and integration service layer, as shown in Figure 3-31.

Presentation Layer

You can integrate AVR frameworks such as ARcore, ARkit, etc. with the DXP core presentation layer, which provides augmentative and virtual reality integration with mobile, web, and desktop applications. AR works on two core applications: marker-based AR application and position-based AR application (also called markerless AR).

CHAPTER 3 DESIGN

- In a marker-based AR application, the image you want to recognize is provided and you know exactly about the thing to be searched using the camera's data (frame). It is like detecting the hard-coded things in your application.

- In a position-based AR application, the image is not available beforehand; you have to recognize and identify features like color, pattern, edge, etc., which exist in the camera frame. In this application, different sensors are required to recognize position and orientation.

Integration Service Layer

After recognizing features and patterns, you need to integrate them with AI-based algorithms along with integration frameworks such as cloud-based integration or ESB-based integration as per your existing application, so that you will be capable of getting data from the existing DXP's application, as shown in Figure 3-31. Once the features and pattern have been extracted by frame, appropriate programming action can be taken to get the objects on the screen using its camera.

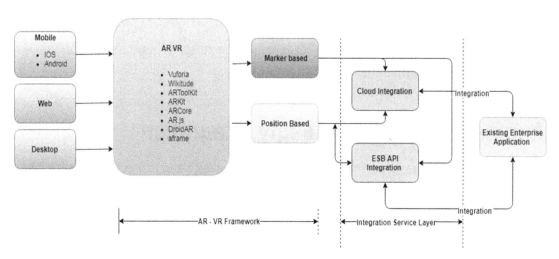

Figure 3-31. *Augmented reality integration*

112

Recent Trends in DevOps

Let's look into the recent trends in DevOps, where applications are built using the containerization approach and deployed on a cluster of nodes using Kubernetes, as shown in Figure 3-32.

Containerization

An end-to-end application can be developed and encapsulated in a single container along with its components such as files, environment variables, libraries, and OS necessary to run the application. The complete set of components in a single container is called an image. The container engine is responsible for deploying these images on hosts. Containers can run inside VMs, physical machines, or public and private cloud. This implies that a host machine can have multiple OS supporting containers that share same physical resources. Docker is the most common and leading containerization system. This approach helps you to scale and increase your storage when it demands.

Features of the container are as follows:

- The required configuration files along with libraries are available in the container.

- Containers are more lightweight than a VM. This makes your application portable, hence it is easily built and deployed.

CHAPTER 3 DESIGN

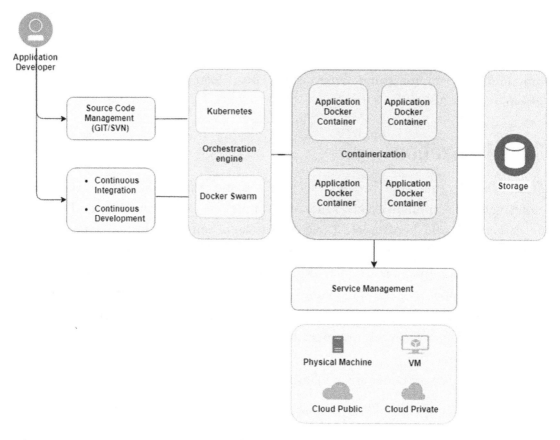

Figure 3-32. *Containerization and DevOps*

DevOps – Continuous Integration (CI), Continuous Deployment (CD)

DevOps consists of the tasks that manages orchestration and cluster management. It also provides features like scalability and load balancing for containerized application.

You develop the application and push the source code in an SCM repository such as SVN or Git, and use CI and CD methodologies to automate the deployment process. You can use Jenkins to build the container images; these images are deployed on multiple Docker clusters using Kubernetes or Docker Swarm.

- *Kubernetes*: Kubernetes is an open-source system for automating deployment, scaling, and management of containerized application. It also distributes the load among containers.

- *Swarm*: Swarm is used for managing a cluster of Docker engines.

Chapter Summary

- We went through different DXP layers and designing of those layers in brief to develop an end-to-end enterprise solution.

- We also went through integration of cutting-edge digital technology like Blockchain, IoT, AI, big data, and AR.

- We went through a variety of open-source frameworks to develop a digital experience platform.

- We looked into the latest trending concept of containerization to develop a solution that can be hosted on any machine, irrespective of OS.

PART II

Development of the Banking Experience Platform

CHAPTER 4

User Interface Design

As we are heading toward an open-source digital experience platform, DXP's provide collaborative user interface. This chapter provides DXP user interface (UI) concepts and shows how to develop intuitive and interactive UI designs. In this chapter we look at:

- The key features of DXP UI
- Architecture and frameworks used in developing DXP UI
- Designing pages and layouts, UI components (such as a widget or port-let), and hooks to integrate with backend services, etc.
- Technology stack to develop a DXP UI
- Case study - Banking experience platform

Let's begin by looking at the key features of a DXP UI.

Key Features

DXP user interfaces are built upon the modern web development approach, using the latest web frameworks and library. Let's delve into the digital experience features and then look into the approach to design and develop the UI.

Simplified Approach

A DXP provides a reusable and intuitive UI. It provides techniques to help create a fancy, and at the same time elegant, UI and enhance the user experience (UX). The object-oriented programming approach makes the components reusable and is developer friendly.

CHAPTER 4 USER INTERFACE DESIGN

Intuitive Architecture

A DXP's information architecture provides content in an organized and intuitive way. Navigating through the application enhances the UX journey. It makes the application simpler and more intuitive to use and the longer visitors stay, the more likely they are going to engage with the content and maximize the chance of buying services or products.

Dashboard

A dashboard is an organized way to provide and present information in an intuitive manner. The dashboard helps in visualizing, tracking, and analyzing data and displays key performance indicators (KPIs), metrics, and key data points to monitor accounts, business processes, department reports, etc. For example, in a banking experience platform, you can analyze account statements and track income expenditures from account statements. Behind the scenes, DXP architecture integrates multiple data sources, for example, files, attachments, services, systems, etc., and provides a single source of data as REST APIs, which integrate with the dashboard and display all data in the form of tables, line charts, bar charts and gauges, etc. A data dashboard is the most efficient way to track multiple data sources because it provides a central location for businesses or users to monitor and analyze data.

Responsive Interface

Grid system layout of a DXP handles the responsiveness of the application, on desktop, mobile, as well as tablets. DXPs work on a mobile-first approach to implement responsive features. The mobile-first approach is the best strategy in the market to make adaptive designs. In the mobile-first approach, content or information is prioritized and sorted into primary, secondary, and tertiary content, for example, the home page should have a company logo and links to products or services. As everything wouldn't fit into a smart phone screen, the DXP provides the ability to prioritize the content as per business requirements. In Figure 4-1, the desktop screen has six UI components: components 1 and 2 are primary content, component 3 is secondary content, and components 4, 5 and 6 are tertiary content. As the resolution of the screen changes, these components start getting wrapped as per the priority of the content, as shown in Figure 4-2. Components 4, 5 and 6 where horizontally aligned, but the mobile view is wrapped up and these components are vertically aligned.

CHAPTER 4   USER INTERFACE DESIGN

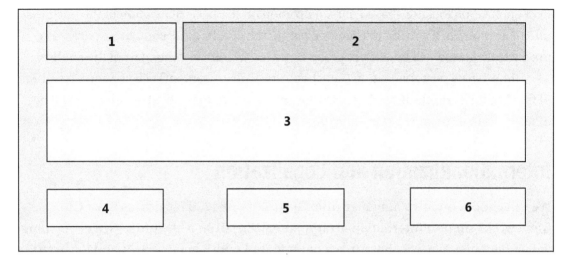

Figure 4-1. *Desktop screen*

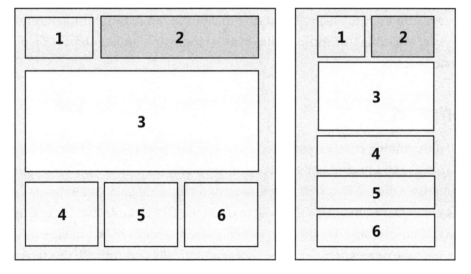

Figure 4-2. *Left: tablet and Right: mobile*

Personalization

Business stakeholders or administrators can provide customizing ability to the user; thereafter the user can also customize the look and feel of the page from predefined templates. A user of the system can decide the content and layout of the application. Business owners or stakeholders can make use of predictive content personalization based on Artificial intelligence (AI)-based predictive algorithms in which similar content, product, and services are displayed on the basis of their previous interaction with application location-based personalization on the basis of location of the business unit. For example, if you log in from one part of the country, the content, services, and products displayed will be different than if you log in from another part of the country. Also, time-based personalization (i.e., theme and advertisement) will change as per time of the day; for example, if you log in at morning time, theme and advertisement will be different than if you log in at evening time.

Internationalization and Localization

We can choose our own language from the list of provided locales, as most UI frameworks support internationalization (i.e. i18n). i18n is the process of developing you application in a way that can accommodated multiple languages and localization (i.e. i10n). i10n is the process of adapting i18n to enable usability in a culture. The process of developing UIs should be done in such a way that they can be localized for language and culture easily. The user can select a language among different languages by selecting the appropriate locale (language), and content is displayed as per selected locale.

Preferences

The DXP presentation component (such as a widget or portlet) provides editable features to make UI integration flexible. For example, REST service call endpoints can be editable from the UI itself. A preference is a key value pair. Preference is stored as metadata, which helps the DXP to make the UI customizable, for example, title, description, etc. are customizable as per business requirement changes. All UI components have names, which can be programmatically determined; this makes presentational component robust in nature.

Integrated Analytics

When you are designing a portal, you need to pay attention to analytics to understand user behavior and patterns. There could be a scenario where you want to track click events and user behavior. Analytics helps you understand the potential customer. Google Analytics makes it easy to record click events. The DXP provides the ways to integrate the analytics framework (e.g., Google analytics, Adobe analytics, etc.).

Search Engine Optimization

Search engine optimization (SEO) is the way to drive traffic. The UI and UX make or break the first experience. If you don't have SEO, it is hard to find your applications on search engines. On the other hand, if you don't have a rich user interface, you don't get interaction and leads. The DXP uses best practice while developing UI to make it SEO friendly. The DXP provides keyword tagging features (i.e., title tagging, metatagging).

User Interface Components

A DXP page typically includes a header with a logo, a navigation menu, presentation component, container area, and footer.

Pages

A web application contains a set of pages that are used to display the application. A page contains different layouts as per the experience requirement, for example, one-column layout, two-column layout, three-column layout, etc.

Layouts

A DXP provides a responsive layout that is used while designing the UX on pages. It is a predefined structured template on which the UI designer can drag and drop presentation components.

- *Navigation layout*: It usually contains three areas: Top (containing logo area and user area), navigation side as shown as container in Figure 4-3, and as Presentation component, as shown in Figure 4-3. The layout is designed based on Bootstrap. There are no restrictions on what layouts or UI (presentation) components are included in any area.
- *Light box layout*: It overlays the page and can be shown and hidden based on click events. It is provided to the user in a sub flow without leaving the current page, enhancing the UX.

CHAPTER 4 USER INTERFACE DESIGN

- *Carousel layout*: It provides the facility for transition between the areas (slides) of a layout, only showing one area at a time. It can be configured for auto play so it loops through the slides once it is loaded.

- *Columns layout*: It provides the basic grid functionality. The layout is built on Bootstrap columns and Cascading Style Sheets (CSS) classes. You can configure the column widths by using CSS classes.

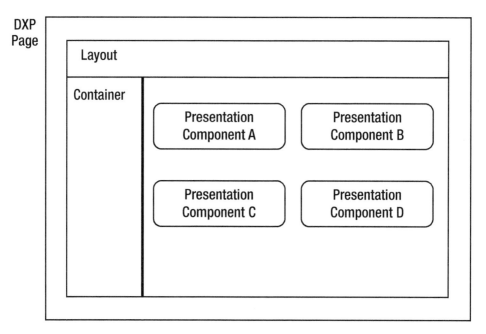

Figure 4-3. User interface components

Navigational Router or Navigation Menu

A consistent navigational router is one of the components that provides users with a sense of orientation and guides them through the application.

Presentation Component

Presentation component (such as a widget or portlet) are independent mini user interface applications usually separated as per use cases or user story. For example, an account summary component to view account details and a bill pay component to pay bill payment of registered payees are two separate presentation components. These components (portlet or widget) have self-sufficient functionality and work independent of each other.

Design Goals

While deigning the presentation component, model-view-controller (MVC) and model-view-viewmodel (MVVM) patterns are used because these architectural patterns provide control over business logic implemented on UI and also provide control over the workflow of the UI application.

The MVC pattern has three components: model, view and controller:

- *Model* will bond with view as well as controller, as shown in Figure 4-4.

- *View* is the user interface that binds the model with the Document Object Model (DOM) and display data to the user, and also enables the user to modify the model.

- *Controllers* are responsible for controlling the flow of the application; if you make a web services request, the controller is responsible for providing a response back to the application.

MVC and MVVM patterns provide the following features to your application and help to achieve goals like upgradable, extendable, lean, and testable:

- Upgradable and extendable presentation components. This can be achieved through API and object-oriented features of the framework. As shown in Figure 4-4, the UI component can extend to the base component; hence, if upgrading the framework, your functionality will not be impacted in the UI component. You can implement common functionality in the base components and other UI components can use it simply by extending the base components.

CHAPTER 4 USER INTERFACE DESIGN

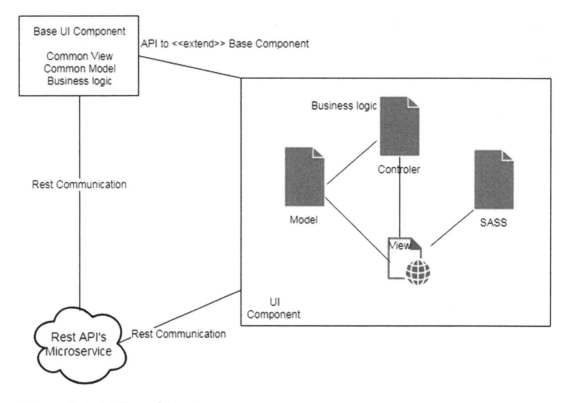

Figure 4-4. *MVC architecture*

- Portlets should be independent of each other, as per digital experience platform business requirement. Angular framework or React library along with Flux library makes a lean-structured presentation layer. Presentation components should be of high quality and should be well tested. Quality is achieved thoroughly and efficiently by testing presentation components using a test-driven approach. Therefore, framework and libraries like Karma, Jasmine, Mocha, Chai, etc., help to achieve high quality.

Communication Between Presentation Components

Sometimes it is vital that one presentation component responds to an action made in another presentation component. For example, after a user has selected a different account from the account list component, the transaction list is updated to show transactions related to that particular selected account in the transaction list component, as shown in Figure 4-5. If presentation components are on the same page, this can be achieved using a broadcast observable design pattern to ensure smooth communication between different components. A broadcast enables you to capture and handle events triggered in one component and actions performed in another component.

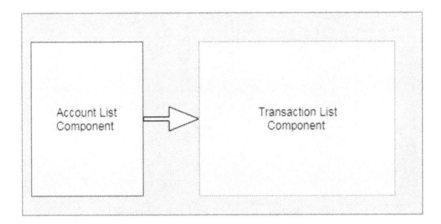

Figure 4-5. Communication between presentation components

Hooks

Hooks are created so that flexibility of work can be maintained between front-end and back-end developers. Hooks are defined services from which back-end integration takes place using REST service calls. A hook itself does no processing—it just calls the hook's implementation, passes the data, and accepts data back from the implementation. This allows exchange of data between back-end and front-end applications. A hook has been defined in the service files to handle data coming from the back-end, and vice-a-versa. When the developer implements a REST service, then the hook data points will be replaced by actual REST service endpoints.

Development Process

A DXP provides a simplified and structured UI development approach. Figure 4-6 outlines the UI development processes.

CHAPTER 4 USER INTERFACE DESIGN

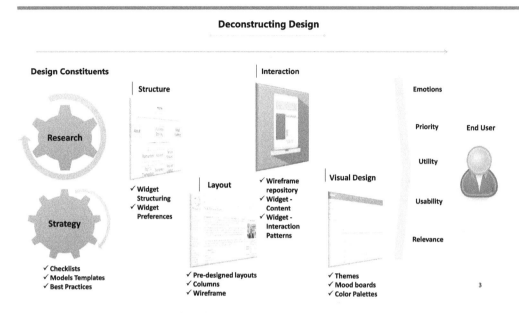

Figure 4-6. *User Interface (Visual) development process*

DXP UI design constitutes of research and strategy of the following process:

1. *Structure*: Build basic UX structure.

   The structure is designed in such a way that it can incorporate the digital strategy of the brand. Contents are classified according to the digital strategy of the organization so that they can provide products and services to users of the application. Templates and models are built to understand the needs of digitization.

2. *Layout*: Select layout on the basis of the UX structure.

   Once a model and templates are selected or built, you need to check and research the business needs to build wireframes for the application. These wireframes can be built using a predefined layout, and these layouts are reusable basic structures built by considering usability by the user.

3. *Interaction*: Classify the content on the basis of the digital strategy of the organization.

   After building the wireframe layout, content is structured in the layout so that it enhances the relevancies of the content. Content is classified in these layouts, which increases the utility of the application.

CHAPTER 4 USER INTERFACE DESIGN

4. *Visual design*: Selection of color coding and arrangement of Visual aids.

   After building layout and classifying content, you need to consider the themes, color coding, color palettes, and mood boards, which are an arrangement of images, materials, and pieces of text that enhance the UX design.

Development Life Cycle

The development life cycle is as follows:

1. Designing:

   Prototype: As shown in Figure 4-7, While designing the User Interface(UI), UI Prototyping and UI Wireframe are essential step towards the success of User Interface Development. While prototyping the UI, you should take care of structuring and layout of the content. Best strategy and processes should be considered (e.g., interactive screen mock-ups as per the mobile-first approach strategy that focuses on the interface and content priority as per mobile screen and then desktop) so that the content should be presented in a meaningful sequence.

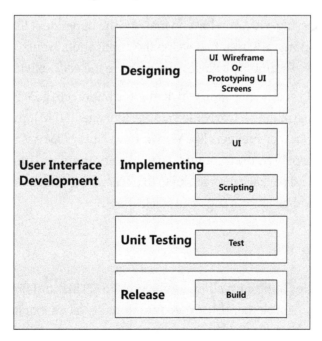

Figure 4-7. User Interface Development life cycle

2. Implementing:

    Construct UI: You construct the UI and presentation components using layout, containers, pages, and UI elements with reference to the wireframe built in the designing phase. These layout, container, and presentation components consist of HTML and JavaScript's (also called ECMAScript).

    Presentation components should be structured and encapsulated in containers and layout from the list of predefined layouts as shown in Figure 4-3.

    Implement Business logic: UI components (view) built are mapped to JavaScript's controllers (controllers) using a model, which helps to control the application and implement bossiness logic on the basic of any change that happens to the model.

3. Testing:

    Unit testing: The developer can test the presentation components with different scenarios or test cases using different testing frameworks (e.g., Karma, Jasmine, Mocha, Chai, etc.).

4. Release:

    While releasing code, you have to manage the dependent libraries and third-party API used to develop the application. Hence build is a process to assemble packages and manage the code efficiently.

    Build: Build your application with a module bundler (e.g., Webpack) and package manager like Node Package Manager (NPM). You can use other package mangers like Yarn or Bower, but NPM has been widely accepted by the front-end developer community. These bundlers and package mangers help in packaging a complex and large scale application's code as a single unit.

Architecture

A DXP is a platform for building a client application in HTML and Java Script (also called ECMAscript). It implements the core functionality as a set of Java Script libraries that you import into your application. It helps in organizing your code into distinct

CHAPTER 4 USER INTERFACE DESIGN

functional modules referred to as presentation components (widget or portlet), which help in managing development of a complex application. This technique lets you take advantage of lazy loading, that is, loading modules on demand in order to minimize the amount of code that needs to be loaded at startup.

UI architecture is built on the MVC or MVVM architectural pattern. MVVM is recognized as a web architectural-based pattern.

- The model represents the data and binds with the view.

- The view is where you represent UI elements, for example, textboxes, buttons, input, etc.

- ViewModel represents the UI-related logic where you can do conditional checks or update certain parts of the web application.

As shown in Figure 4-8, the model will bind with View (HTML template) and ViewModel (components' scripts). Events fired in the view can be recognized in a components' scripts and any change in a components' script will be reflected in the view or vice versa. This will help you to maintain the state of the application; in this case you have control over your business logic on UI.

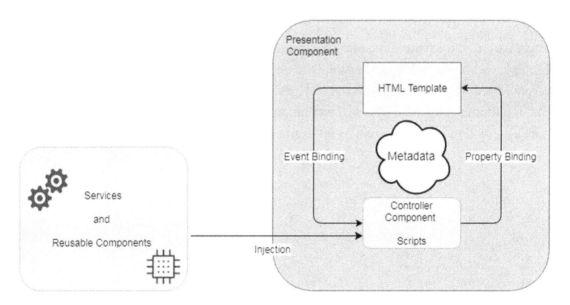

Figure 4-8. *Component architecture*

131

Reusable data or logics can be separate reusable components, which can be shared by injecting into presentation components; this can be achieved using dependency injection. Angular uses a module loader to load all components, modules, and services, rather than explicitly putting script tags into the UI (presentation) component's template HTML. The presentation component only needs to know about its immediate dependent libraries. Dependency of imported libraries are loaded automatically.

The presentation component consists of HTML (templates) and controller components. Data binding between controller and HTML enables you to synchronize application state (model) and the view. In the case of unidirectional data binding, any change in the state of the application updates the view; and in the case of two-way data binding, it binds properties and events together as a single entity so that any change in the model updates the view and vice versa. As shown in Figure 4-8, event binding makes your application respond to user input by updating the application model and data associated with the model. Property binding interpolates values that are computed from the application model into the view (HTML)

DXP UI Technology Stack

DXP technology stacks are a combination of different frameworks and programming languages used to create a flexible, responsive UI that is mobile and desktop compatible. Multiple technology stacks are available to implement the DXP's presentation layer. Each layer of the application builds on the features of the one below it. Figure 4-9 shows the major building blocks of a DXP's UI technology stack, and you can add other custom packages using a package manager, for example, NPM, Bower, etc.

CHAPTER 4 USER INTERFACE DESIGN

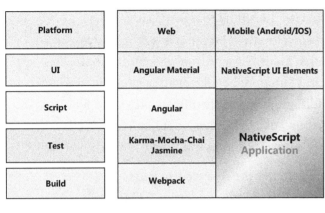
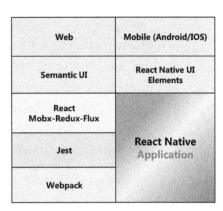

Figure 4-9. *Technology stack*

Presentation components are built on well-known, proven standards and technology stacks. Let's begin by looking at the Angular technology stack to implement the DXP's presentation layer.

Angular Technology Stack

The Angular technology stack (ATS) consists of multiple frameworks and libraries (Angular Core, Angular Material UI, Bootstrap, Swagger, Jasmine, Webpack, etc.).

CHAPTER 4   USER INTERFACE DESIGN

Angular Core

The Angular framework is leveraged to provide many features to speed up UI development:

- *Cross platform*: Angular provides the ability to reuse your code to build an application for any development target. It is a progressive web application approach that loads like a normal web application but also provides features and functionality like push notification and device hardware access traditionally available only to native mobile applications. It has the capability of a hybrid web application, which works on desktops and mobile devices across multiple browsers.

- *Development friendly*: Angular serves the first view of your application on the Node.js server, instantly rendering HTML while developing the application.

- *Code splitting*: Angular loads scripts quickly with the router component, which delivers automatic code splitting so users only load code required to render the view the user has requested.

- *Productivity*: The Angular command-line interface (CLI) tool quickly creates UI views with simple and powerful templates.

- *Test-driven approach*: With a library like Karma and a framework like Jasmine, you can know if you've broken anything every time you save the code while developing the application.

- *Server-side rendering*: Angular provides server-side rendering by using Angular Universal, a technology that runs applications on the server. Angular Universal generates static application pages on the server through a process called server-side rendering (SSR). It enables the web crawler to index your application and optimize your application so it will be easily searchable, linkable, and navigable for web crawlers.

- *Angular APIs*: Angular has an extensive set of APIs that are flexible and customizable. When we say API, it does not mean REST APIs. API refers to any programming interface. For example, `https://angular.io/api`

Angular Support Library

The Angular support library includes the following:

Material UI

Angular Material is Bootstrap components written in Angular by using Google's Material Design specification. All the UI (presentation) components such as accordion, table, gauge, charts, etc. have been split into separate importable modules, which are reusable across applications. As shown in Figure 4-9, Angular Material provides UI capability to the ATS.

Bootstrap

Twitter Bootstrap is used while developing the UI (presentation) component of a DXP to speed up development. However, a DXP is lean and flexible, hence the entire template HTML is customizable. It is possible to use another CSS framework as per business requirement. It works on the mobile-first approach to implement responsive features.

SASS (Syntactically Awesome Style Sheets) – CSS Preprocessor

SASS provides a simpler, more elegant, syntax for CSS and implements various features that are useful for creating and managing CSS, such as nested rules, variables, mixins, selector inheritance, and many more. It also helps to keep everything organized and allows you to create style sheets faster.

Swagger

Swagger helps you to mock the web services. The UI (presentation) component has service data hooks that communicate with a server through a REST API. All the REST APIs are defined in Swagger. Code generated by Swagger provides a simple and well-defined interface to the REST APIs. This enables fast development, as the developer can see from the data module documentation exactly which methods are available, what parameters are accepted, and the JSON in which the data is returned.

NativeScript

NativeScript creates the native application iOS and Android Apps with Angular. It provides the abstractions needed to access the underlying native platforms; for example, it provides a JavaScript API that translates application JavaScript code into native (iOS or Android) gestures API calls. It provides modules to access native device and platform capabilities. As shown in Figure 4-9, it provides Native Mobile application building capability to the angular Technology stack.

Karma-Mocha-Chai

Testing a web application is not as simple as testing a back-end application because you have to test front-end code on multiple browsers and their versions. Karm, Mocha, and Chai help you to test your code on multiple browsers. Karma runs the test, whereas Mocha and Chai are used to write the test.

Karma allows you to test your code on browsers and devices; it starts the browser and runs the test on it. Chai and Mocha provide an assertion library that can be integrated with any JavaScript testing framework. They provide testing capability to the ATS.

Jasmine

Jasmine is a behavior-driven development framework for testing Java Script code. It is used to test behavior of the functionality written in JavaScript. It has simple syntax so that you can easily write test cases.

Webpack

Webpack is a static module bundler. When Webpack processes your application, it internally builds a dependency graph that maps every module yours project needs and generates one or more bundles. It provides application-building capability to both the Angular technology stack (ATS) and React technology stack (RTS).

Gulp

Gulp is used as a default build tool for UI (presentation) and themes development. It helps in automating build tasks like building CSS, HTML, and ECMAScript along with other tasks like minification, watching changes in source code, linting for errors, etc.

NPM

DXP UI (presentation) component development relies on NPM (Node Package Manager) so that you can incorporate other NPM packages into your development process. It provides package management capabilities to both the ATS and RTS.

React Technology Stack

The RTS consists of multiple libraries such as React, Semantic UI, React Native, Redux, MobX, and Flux. You can make your own framework using these libraries.

React

React is a component-based JavaScript library for building a UI that deals with the view in the MVC. The React library is leveraged to provide many features to build a dynamic user interface:

- *Server-side rendering*: Next.js is the framework for the server-side rendering of a React-based application. It provides a flexible way to completely or partially render your application and optimize it so it will be easily searchable, linkable, and navigable for web crawlers.

- *Performance*: Virtual DOM in React makes the UX better and it works faster.

- *Reusable component*: This component has its own logic and controls its own rendering, and can be reused wherever you need. Code re-se helps to make your apps easier to develop and easier to maintain.

React Support Library

The React support library includes the following:

Elemental UI or Semantic UI

Semantic UI is React's official integration CSS framework that helps create beautiful, responsive layouts using HTML. It uses simple phrases called behaviors that trigger functionality. As shown in Figure 4-9, Semantic UI provides UI capability to the RTS, but you can use other CSS frameworks (e.g., Elemental UI, which is the UI toolkit for a React-based application).

React Native

React Native is a platform for creating a native mobile application using React. It provides a set of React components that bind to their native mobile counterpart; it also provides features to create your own components and bind them to native mobile code. As shown in Figure 4-9, it provides native mobile application-building capability to the RTS.

Redux-MobX

Redux is a simple state management engine for JavaScript. Redux helps you to write applications that behave consistently, and run in different environments (client, server, and native). MobX is a simple, scalable state management solution. MobX is just a library to solve state management problems, not an architecture or even state container in itself.

MobX is used for small-scale project, whereas Redux is mainly used for complex and large-scale projects. MobX has more than one store for data storage, whereas Redux has only one large store for data storage. Redux and MobX both are the libraries that are used to manage the application state in one way or the other. These libraries are mainly combined with front-end libraries like React to develop the UIs more interactively and to show changing data over time.

Flux

Flux is the application architecture that Facebook uses for building client-side web applications; it's a pattern rather than a formal framework. It is a kind of architecture that complements React and the concept of unidirectional data flow. It is the data layer in JavaScript applications and building client-side web applications. React takes care of V or the view part in MVC, whereas Flux is a programming pattern that takes care of the M in MVC.

Jest

Jest is used by Facebook to test all JavaScript code including React applications. As shown in Figure 4-9, it provides testing capability to the RTS.

Evaluating UI frameworks

To evaluate UI frameworks, you should consider the following:

Data Flow

The main difference between Angular and React is the way of handling data and managing the state of application. Angular is a fully featured MVC framework. React is just more of a 'V' in the MVC. Angular allows two-way data binding, while React allows one-way data binding. Unidirectional data flow, also called one-way data binding, means any changes you make to the model affect the view, but not the other way around. This way, the data only flows in one direction, whereas with two-way or bidirectional data binding any changes you make to the model affect the view, and vice versa; hence with React, state management is provided and managed by a third-party library (Flux, Redux, MobX). Angular is capable of managing state itself, but React needs to integrate with other third-party libraries. Angular has more features out of the box than React.

Language

Angular is a JS framework build using typescripts, whereas React is a JavaScript library but recommends using JSX(XML syntax to JavaScript). Instead of writing the traditional way—a classical approach of separating markup (HTML) and logic (JS)—React combines them in the components using an XML-like language that allows you to write markup directly in your JavaScript code.

Performance

DOM is the Data Object Model of the DXP application. Angular uses the browser's DOM, while React uses a virtual DOM. A virtual DOM is a simplified version; therefore, by using a virtual DOM you can change any element very quickly without rendering the whole DOM. Therefore, React has better performance over Angular.

Both technology stacks are flexible and powerful. Their usage depends upon the business application. Both accomplish the same thing but React needs the support of an additional library that provides framework capabilities to the RTS; otherwise it is ideal for a logicless application.

CHAPTER 4 USER INTERFACE DESIGN

Best Practice

A UI should be perceivable, operable, understandable, and robust. You should keep the main menu structure simple and consistent across layouts, and make sure it's intuitive and easy to use. While keeping the layout simple, also make sure different elements are easily identifiable as primary buttons, secondary buttons, action items, or menu. You should group menu navigation based on user needs and mental model. Organize content into relevant groups and categories to increase its relevance:

Perceivable:

- You should review all color and contrast settings.
- You should check alternative text applied to all nontext content.
- You should provide a media alternative and description to each and every component.
- Your content should be presented in a meaningful sequence.
- Color is not the only visual means for conveying information; you should also check the shape, size, and content, and include animation.

Operable:

- Ensure navigation is consistent across pages and layouts, making users' navigation easy. Match the navigation flow with the user mental model.
- Content should be operable through a keyboard interface.
- Check for missing headings and blank labels.
- The purpose of each link should be determined from the link text alone.
- Ensure focusable components receive focus in a meaningful order and a focus indicator is visible.

Understandable:

- Consistent navigation should be within a set of web pages.
- Changing the setting of a UI component should not automatically cause a change of context.

CHAPTER 4   USER INTERFACE DESIGN

- Input errors should be identified, and suggestions should be clearly described and provided to the user in text.

- Rewritten site content to a lower reading level.

- Provide customized content, based on the user's product and web application usage patterns.

- Ensure the error messages are easy to understand, and display the solution to the problem.

Robust:

- All UI components should have names, which can be programmatically determined.

BXP – Case Study

A banking experience platform is the technology-driven platform that links multiple technologies into one. A BXP solution is more about optimizing, rebuilding, and connecting multiple platforms.

Consistency Across Locations

BXP features like localization (i.e., l10n) and internationalization (i.e., i18n) provide one front end that is applicable to all the countries, thereby providing consistency across different regions and countries. The current banking application has multiple screens to support functionality for bill payments and managing payees where there are separate and multiple interactions (screen) for money movement workflow. The BXP provides a solution where there is a single logical user flow for any kind of money movements.

Consistency Across Application

One of the key aspects of the UX design process is to ensure consistency from a usability and design perspective throughout the application. In the existing application, setting or editing addresses, e-mails, and phone numbers followed different patterns and hence wasn't intuitive to users. The BXP provides a consistent design approach for all the scenarios.

CHAPTER 4　USER INTERFACE DESIGN

Unified and Collaborative Approach

The BXP provides a single platform for retail banking customers as well as business banking customers. Hence, it reduces the bank's operational cost to maintain different applications for different types of users. The BXP enables customers to define smart actions that help them easily automate manual tasks.

BXP UI Layouts/Containers

Layouts or containers are used to create structure for presentation components:

- *Navigation layout*: It usually contains three areas: top (containing logo area and user area), side, and content, as shown in Figure 4-10. The layout is designed based on Bootstrap. There are no restrictions on what layouts or UI (presentation) components are included in any area.

- *Columns layout*: It provides the basic grid functionality. The layout is built on Bootstrap columns CSS classes. You can configure the column widths by using CSS classes.

BXP Dashboard

The BXP dashboard is a user interface that organizes and presents information in an intuitive and interactive manner. The dashboard visually tracks, analyzes, and displays all linked accounts and transactions to monitor accounts and manage wealth. For example, in a BXP application you can analyze account statements and transactions by filtering transactions on the dashboard. Behind the scenes, a dashboard connects to multiple data sources and APIs, but on the surface displays all this data in the form of tables, line charts, bar charts, and gauges. A data dashboard is the most efficient way to track multiple accounts and transactions, because it provides a central location for retail as well as business banking users to monitor and analyze their wealth.

Problem: XYZ bank wants a responsive design to display accounts and transactions in the user's dashboard.

CHAPTER 4   USER INTERFACE DESIGN

Solution: The BXP provides an extensive set of layout to model presentation components. As shown in Figures 4-10 and 4-11, we have used navigation layout along with two columns structures that contain the account summary presentation component and transaction presentation component.

The Account Summary UI component displays various accounts one holds with XYZ bank along with summarized details.

The transaction UI component displays transactions associated with the account.

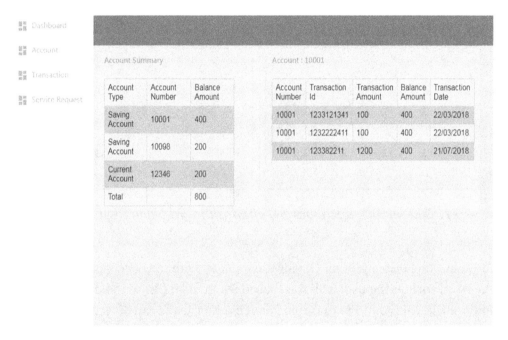

Figure 4-10. *BXP dashboard*

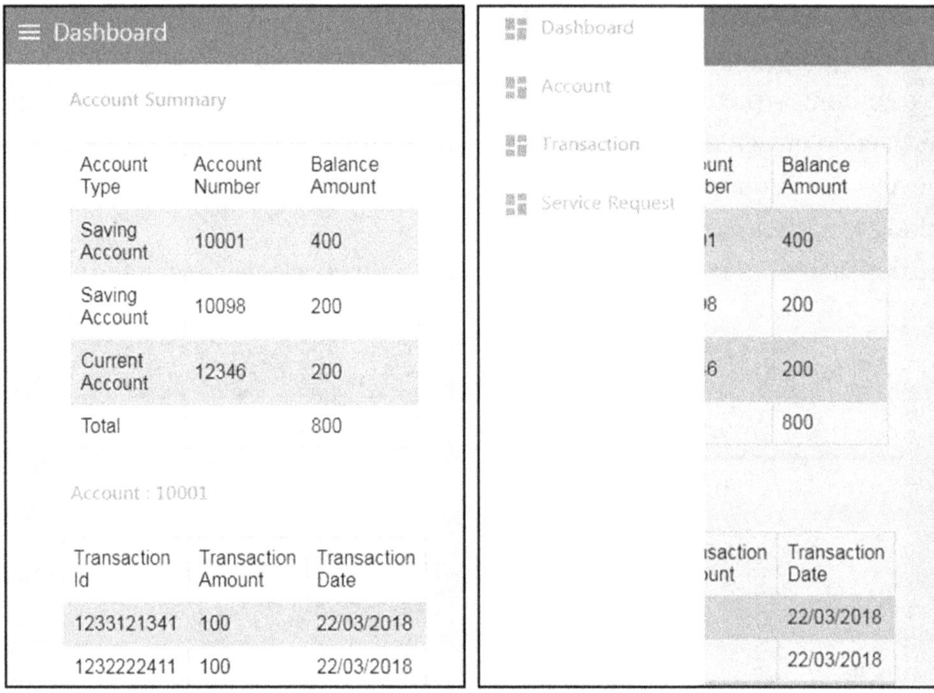

CHAPTER 4 USER INTERFACE DESIGN

***Figure 4-11.** BXP mobile dashboard (Left: dashboard view; Right: navigation view)*

The side panel displays other UI views associated with the BXP application.

The BXP provides the user with role-based featured UI (Presentation) components, as mentioned in Table 4-1. For example, retail banking users can only access those components with access to the retail banking group. Business banking users can access the business banking group's components.

Table 4-1. *Role-Based Components*

User or Role	UI Components
Retail banking UI components	Login component, profile component, account summary component, payments component, transactions component, manage payee component
Business banking UI components	Login component, profile component, account summary component, payments component, transactions component, authorizations, batch upload component, manage payment orders component, draft payment orders component, manage payee component
Wealth management UI components	Portfolio details component, portfolio summary component, portfolio transactions, portfolio performance valuation component
Bill payment UI components	Third-party integration UI component, mobile recharge UI component, manage biller UI component
E commerce UI components	Add biller's component, manage biller's component

BXP UI (presentation) retail banking components, which enhance user experiences, are mentioned in Table 4-2.

Table 4-2. BXP UI Components

UI Components	Functionality	Features
Login	User to be authenticated with username and password.	User is authenticated using username and password. User is taken to specific page after logging in.
Profile	Display read-only information about the logged-in user.	It displays personal information about the end user.
Account summary	Display summary of user's account, credit card, debit cards.	It displays aggregated balances of all accounts, debit cards, credit cards.
Transactions	Display overview of user's selected account transactions.	User is able to see most recent transaction for particular selected account. User is able to load transactions incrementally. User is able to filter credit or debit transactions. User is able to search transactions. If the number of applicable transactions is greater than the value defined for display in the components' preferences, then by using lazy loading functionality more transactions will be fetched from back end and be displayed under the list.
Manage payee	User is able to add new payee and edit existing payees.	It provides ability for the user to create, search, edit manual payee.
Manage payment orders	User is able to view the payments orders applicable to the current user's entity.	It provides ability to load payment orders incrementally and access all payment orders through indexed pages. User is able to view and edit scheduled payment order, etc.

The BXP provides digital banking capability to move toward a DXP. It provides capability for building and managing a banking application using open-source technology. You can create a personalized dashboard based on the modular and flexible structure of a DXP. This will help to create a personalized UX. It helps banks to optimize and revolutionize their business processes, for instance, it provides easy registration process ability by integrating with KYC (know-your-customer) services, which enhances the registration process.

Chapter Summary

- The testing framework helps to analyze the code as per functional requirement and enhance usability. Therefore a DXP is intended to highlight the strategy that is currently being used for creating enriched and intuitive UI design.

- The DXP technology stack plays a vital role while designing the user interface. It provides information architecture and it helps in prototyping and keeping the main menu structure simple and consistent across layouts, making sure it's intuitive and easy to use. Layout ensures consistent layout with proper same or similar content categorization for finding or searching. While keeping the layout simple, also make sure different elements are easily identifiable.

- Mobility ensures a website, when used across devices and platforms, gives similar or same experiences including mobile apps and where possible makes it easy for customers to continue with their actions across platforms. Content provides customized content, based on users' product and website usage patterns; it ensure the error messages are easy to understand and display a solution to the problem.

- A DXP also ensures context-based, relevant information is displayed at the appropriate places. Navigation is consistent across pages and layouts, making user navigation easy. As shown in Figures 4-10 and Figure 4-11, the Transaction component has fewer fields in mobile view compared with desktop view. In mobile view, less important information would wrap up.

- DXP concepts provide the comprehensive view to develop next-generation digital applications that are pertinent across all domains.

CHAPTER 5

Designing the Integration Layer

After going through extensive UI concepts, now we look into the integration layer, which helps us to integrate with the DXP UI and further enhance DXP capabilities with a lean and flexible integration platform. Existing systems are integrated with the maintainable and scalable integration layer.

The key features of the DXP integration layer are the following:

- Various types of integration platform.
- Architecture and frameworks used in integration layers.
- Technology stack to develop integration layer.
- Case study – banking experience platform.

At the outset, we take into the picture the introduction and features of the DXP integration layer.

DXP integration concepts help you make effective decisions on Web API (application program interface) management tools to integrate services with the DXP UI. According to the business requirement, the API management tool can be used and integrated with the DXP. Business requirements can help you to focus on understanding the basic requirements of the integration, for example, some requirements focus on portal and some on analytics services, so decisions should be made as per the requirement.

CHAPTER 5   DESIGNING THE INTEGRATION LAYER

DXP integration concepts will address frequently asked questions in Integration Style, Integration System and services Sections of this Chapter on integration platform, such as the following:

- Will the business requirement focus on management of existing services using API gateways?

- Will the business requirement focus on REST (Representational State Transfer) services or will the requirement need to use other service protocols such as Simple Object Access Protocol (SOAP) or Java Message Service (JMS)?

- Will the business requirement need flexible configuration, routing options, and user management (authorization) using different authentication and security standards (for example, Open Authorization (Oauth), Lightweight Directory Access Protocol (LDAP), Security Assertion Markup Language (SAML), Kerberos, etc.?

- Will the business requirement need a caching mechanism?

- Will the business requirement need event-driven architecture or synchronous HTTP calls?

- Will the business requirement need an API management solution on premise or on the cloud platform?

Integration Consideration

A DXP takes into account the current environmental factors, which are systems already available in the environment of the organization such as customer relationship management (CRM), enterprise resource planning (ERP), other services, databases, content management system (CMS), rules engine, and solution gaps; and coexistence expectation, which tries to address the collaborated solution for the organization in the most cost-effective way. Hence while designing the digital platform, the approach is to design an integration layer that covers below mentioned points:

- *Minimize the risk of transition*: The integration layer should be loosely coupled so that if any kind of transition or migration happens in any application in the DXP, it would not impact other services or applications.

- *Maximize business value*: Microservices over monolithic service architecture will provide maximum business value by breaking down functionality to the most basic level and then abstracting the related services.

- *Lower the total cost of ownership and management*: Microservices architecture will increase the decoupling and separation of concern, hence the code base would be easier to manage as each service in the application would be independent of other services and thus it would be easier to add new feature or functionality to the platform. Conversely, in monolithic architecture, adding a new functionality will require smoke testing of the whole application.

- *Interface details*: Go through the interface details, for example, API's operations, inputs, outputs, and underlying types such as XML or JSON, which helps you to integrate the existing or third-party APIs with the DXP application.

- *Data requirements*: A DXP consumes data in any format such as XML or JSON using any service protocol such as RESTLESS-SOAP or JMS, but interacts in RESTFUL-JSON within an application. RESTFUL-JSON interaction makes the application's integration layer flexible and lightweight. REST services are architectural style. REST uses HTTP protocol and HTTP methods like GET, POST, PUT, and DELETE to communicate between client (Angular application) and server (integration) application; whereas, as SOAP is a message exchange style between client and server, SOAP services are Remote Procedural Call (RPC) architecture style, which would have service metadata (i.e., contracts) to communicate between client and server.

  - *Restless*: These services work with resources as well as its operations such as PUT, POST, DELETE etc. You need to identify the services and operations attached to REST Service, for example, operations such as SaveTransation(), GetTransaction(), DeleteTransaction(), UpdateTransaction() on transaction services. A service contract has to be shared with the client; the client will use these details to call the services.

CHAPTER 5 DESIGNING THE INTEGRATION LAYER

- *Restful*: These services work with resources instead of operations. Communication between client and server happens using a Unified Resource Identifier (URI) over HTTP protocol using HTTP methods such as GET whenever someone needs to get the representation of an existing resource. PUT is used to add a new resource into the system. POST is to modify the existing resource, and DELETE is to remove the resource from the system.

- While calling SOAP services from the client, the dispatcher in the web services would first deserialize the SOAP message, and then identify the operation from the message to be performed. Actions are mapped with the service methods, but while calling Restful services you have to identify the resources like a transaction, then the HTTP method (GET, PUT, POST, and DELETE) will identify the method to be called. Each method in web services is mapped with the HTTP method.

- *Integrating methods*: Integration methods are decided on the basis of available interface and requirements of the integration patterns. We further explore integration methods in detail in this chapter.

- *Security*: API layer security would raise the accessibility of the service calls. You need to ensure that only authenticated users of the application can access the API. It will differ on the basic of architectural pattern.

  - *Monolithic architecture*: In this architecture, the entire application is a process; the security module is implemented to provide authentication and authorization to the user. When a user logs in to the application, the security module of the application authenticates the user. After verifying the user details, a session is created for the user and a unique session ID is provided with the session. The session stores login user information such as name, permissions, and roles. The server is responsible for managing the session, and sends the session ID to the client and back to the server in subsequent requests. This session ID would be used to verify the user details.

- *Microservices architecture*: In this architecture, the application is split into multiple microservice processes, and each process holds the business logic of that particular module. Each microservice needs to be authenticated and authorized; hence, the logic of authentication and authorization wouldn't be implemented in every microservice. To overcome this issue, a client application such as an Angular application can access the services through an API gateway where each of the services are registered and controlled. Authentication and authorization modules would be implemented in the API gateway, and all the microservices are registered with the API gateway. The API gateway is exposed to the client application and is responsible for routing requests to the appropriate microservices after verifying the user details.

- *Legacy modernization*: You need to check whether the system has to interact with the legacy system, because the integration method and interface will differ and should be considered while designing the application.

Data Formats

Data formats the integration layer. There are mainly three types of data format, flat, relational, and hierarchal, but it is recommended to use the hierarchal data format because of its efficient structure.

- *Flat*: It has one record per row, but there is a chance of duplicate entry in these things. The flat data format is not recommended in an integration system, but it is used in file systems and databases because it contains additional payload. In the following example, Account ID and Name are redundant in every row for a particular account ID, hence increasing the payload while transporting data in the network and therefore increasing the bandwidth.

Account ID	Account Name	Transaction Type	Transaction Amount	Currency
ABCBANK1234	SourabhhSethii	NEFT	1250	INR
ABCBANK1234	SourabhhSethii	RTGS	850	USD

CHAPTER 5   DESIGNING THE INTEGRATION LAYER

- *Relational*: This data format is used in Soap protocol to pass payloads that contain relational data in the SOAP envelop in XML format, such as account number and transactions. In the following example, account tags hold transactions in it.

```
<Account>
        <Id>ABCBANK1234</Id>
        <Name>Sourabhh Sethii</Name>
        <Transaction>
                <Type>NEFT</Type>
                <Amount>1250</Amount>
                <Currency>INR</ Currency >
        </ Transaction >
        <Transaction >
                <Type>RTGS</Type>
                <Amount >850</Amount>
                <Currency>USD</ Currency >
        </Transaction >
</Account>
```

- Hierarchal: Hierarchal structure provides a lightweight and efficient structure to API providers to communicate between different systems, of which Restful-JSON data hooks are the best example, which helps us to quickly create and solve complex data transformation issues.

```
{
        "account": "ABCBANK1234",
"transaction": [{
                        "Type": "NEFT",
                        "Amount": "1250",
                        "Currency": "INR"
                },
```

154

CHAPTER 5 DESIGNING THE INTEGRATION LAYER

```
                    {
                        "Type": "RTGS",
                        "Amount": "850",
                        "Currency": "USD"
                    }
                ]
            }
```

Integration Services

In today's technological world, there is a need to integrate multiple systems with the DXP. There are multiple platforms and design patterns to support in-built capability and pluggable features, to support DXPs and digital solutions where you integrate these services with the UI layer without changing the existing systems. This further enhances the capability and flexibility of the DXP.

A DXP enhances the digital capability by supporting various integration services and styles that cover requirements from small-scale business to large-scale business, as shown in Figure 5-1.

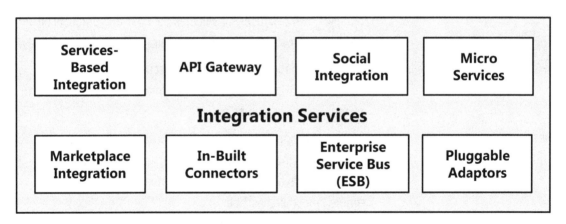

Figure 5-1. *Integration services*

Integration services supported by DXP and its capabilities are mentioned in Table 5-1.

Table 5-1. Integration Services and Capabilities

Integration Services	Capabilities
Services-based Integration	Restful and Restless (SOAP) service integration provides digital integration capabilities. It has small-scale to large-scale application data delivery and integration capabilities.
API gateway integration	An API gateway is an API delivery-based application system used to interact with multiple applications. DXP integration with an API gateway system enhances secure and flexible data delivery, and integration with existing applications.
Social integration	The DXP integrates with social collaboration platforms such as, Facebook, Twitter, Instagram, and many more. It consumes their API and integrates the data in the DXP's application.
Microservices integration	Microservices are designed in such a way that every individual service is independently deployable, small and modular services in which services run as a unique process and communicate and deliver data efficiently between the multiple systems to serve a business goal.
Marketplace integration	Marketplace integration helps business users to integrate with multiple channels: to buy or sell their services and products on different channels while consuming a single service.
ESB (enterprise service bus)	ESB is used in medium-scale to large-scale business requirements where multiple systems inter with each other, for example, a banking domain.
In-built Connectors	An in-built connector has capabilities to connect a DXP's application to multiple components, such as a JDBC connector used to access a database or GRPC connectors used to do remote procedure calls, etc.
Pluggable adapters	Integration frameworks have various pluggable adaptors such as JMS adapter, data stream adaptors, etc. These adapters are used to access and convert data from one form to another form. You can access data streams using these adaptors.

Integration Styles, Protocols, Systems, and Patterns

A DXP supports multiple design patterns and integration platforms as mentioned previously. Integration patterns supported and their implementation models are described in the consecutive segments of this chapter.

Integration Styles

There are different types of styles to consume data, for example, RPCs (remote procedure calls) and messaging file transfer, database, RMI (remote method invocation), that help to integrate multiple applications so that they can exchange data or information with each other.

Remote procedure calls

- *gRPC* (remote procedure calls): gRPC is a modern open-source high-performance RPC framework that can run in any environment. It can efficiently connect services in and across data centers, with pluggable support for load balancing, tracing, health checking, and authentication. It is also capable of browsers to back-end services connecting devices and mobile applications. Hyperledger fabric blockchain project uses gRPC to communicate to the blockchain network.

Messaging File Transfer

- *JMS*: It is used to send messages between applications; it is asynchronous in nature, which means the client is not required to send a request, and the message will arrive automatically to the client. It is of two types. One is the point-to-point messaging domain, where one message is delivered to one receiver only and a queue is used to achieve that. The other is the publisher-subscriber messaging domain, which is like broadcasting in that one message is delivered to all the subscribers; to achieve that, a destination called a topic is used to hold and deliver messages.

- MQTT (Message Queuing Telemetry Transport): MQTT is a publisher-subscriber based messaging protocol that works on top of the (Transmission Control Protocol/Internet Protocol) TCP/IP protocol. It is designed for constrained devices with low bandwidth. It is best suited for IoT-based applications, as it allows you to send commands to controls, and to read and publish from IoT sensors.

Database

- The application stores data in a database; you can integrate databases and consumes the shared data. You can integrate your application to SQL as well as NoSQL databases as per use case requirement.

File Transfer

- You can consume data from the files. Applications produce files of shared data for other applications to consume, and consume files that others have produced.

Integration Protocols

Integration protocols (also called web services protocols) define the structure and definition of message transfer between two applications.

SOAP

It is a messaging protocol exchanging information while implementing web services. It uses the XML message format and works with HTTP protocol for message transmission. The message format mainly contains three elements: envelop, header, and body. An example of a typical SOAP envelope would be the following:

```
POST /Transaction HTTP/1.1
Host: www.example.org
Content-Type: application/soap+xml; charset=utf-8
Content-Length: 200
SOAPAction: "http://www.w3.org/2003/05/soap-envelope"
```

```
<?xml version="1.0"?>
<soap:Envelope xmlns:soap="http://www.w3.org/2003/05/soap-envelope"
xmlns:m="http://www.example.org">
<soap:Header>
</soap:Header>
<soap:Body>
<m:Balance>
<m:Amount>2000 </m:Amount>
</m:Balance>
```

```
</soap:Body>
</soap:Envelope>
```

XML_RPC

XML_RPC is an RPC protocol that uses the XML format to encapsulate the message and send it over HTTP protocol. An example of a typical XML-RPC request is:

```
<?xml version="1.0"?>
<methodCall>
<methodName>account.getBalance</methodName>
<params>
<param>
<value><i4>Sourabh_Sethi</i4></value>
</param>
</params>
</methodCall>
```

An example of a typical XML-RPC response is:

```
<?xml version="1.0"?>
<methodResponse>
<params>
<param>
<value><string>2000 USD</string></value>
</param>
</params>
</methodResponse>
```

JSON-RPC

It is an RPC encoded in JSON format, as shown in the following example. It is a simple and lightweight protocol, and the DXP can consume data in any format but expose and interact in this protocol with other applications.

An example of a typical JSON-RPC request would be:

```
{
        "account": "ABCBANK1234",
}
```

CHAPTER 5 DESIGNING THE INTEGRATION LAYER

An example of a typical JSON-RPC response is:

```
{
        "account": "ABCBANK1234",
        "transaction": [{
                        "Type": "NEFT",
                        "Amount": "1250",
                        "Currency": "INR"
                },
                {
                        "Type": "RTGS",
                        "Amount": "850",
                        "Currency": "USD"
                }
        ]
}
```

JSON-WSP

It is same as the JSON-RPC protocol but has a service description specification with documentation method, name, and description provided along with requested details, as shown in the following example.

An example of a typical JSON-WSP request is:

```
{
    "type": "jsonwsp/request",
    "version": "1.0",
    "methodname": "getTransactions",
    "args": {
        "account": "Sourabh_Sethi"
    }
}
```

An example of a typical JSON-WSP response is:

```
{
    "type": "jsonwsp/response",
    "version": "1.0",
    "servicename": "Transaction",
    "methodname": "getTransaction",
    "result": [{
        "username": "Sourabh_sethi",
        "transaction_id": 123456,
        "amount": "200",
        "type": "credit"
    }, {
        "username": "Sourabh_sethi",
        "transaction_id": 123457,
        "amount": "1200",
        "type": "debit"
    }]
}
```

Integration Systems

Integration systems consist of messages and their transformation from one form to another. We look into messaging systems, their construction, and transformation in this section.

Messaging Systems

A messaging system defines the message format and its transformation capabilities, where messaging channels are defined as per the service that contains the exchange in a piece of information.

Pipe, filter, and routers perform the complex processing on a message or exchange (also called a piece of information), whereas filtered and processed data is passed to other messaging system with the help of routers.

Messaging endpoints are the connections to a messaging channel to send and receive messages.

Message Routing

Message routing is used to handle the scenarios where a single logical service is interacting with multiple existing systems using the list of dynamically specified recipients.

Message routing is used along with a splitter, where one processes the exchange if it contains multiple recipients, each of which may have to be processed in a different way; after processing, one combines the results of individual related exchanges using an aggregator.

Message Construction and Transformation

Event messaging is used to transfer events from one application to another application, which is identified by a correlation identifier that contains a return address and is handled by a request reply message.

Messages are wrapped in an envelope so that the existing system participating in a messaging system can establish a secure message header encryption method.

Content filter and claim check patterns deal with the volumes of data, where large messages are handled but one is interested only in few data items from the entire message.

Integration Patterns

Integration patterns are the solution to commonly occurring integration problems, which are mainly divided as channel pattern (how messages are transported across channels), routing pattern (how messages routed between sender and receiver), transformation pattern (transformation of messages as per sender and receiver), and endpoints (how messages are consumed and exposed).

Pattern – Simple (Internal) Integration

The *publisher-subscriber design pattern* ensures that multiple applications consume the message at a time. Information standardization is based on source and synchronization of target systems with data required for its processing. The main focus is to synchronize information based on business events, thus removing any processing delays and overcoming the pain of managing a batch window.

The *proxy design pattern* provides a single point of entry to the target application; in addition, it supports policy implementation and ensures compliance monitoring. It reduces time of solution, price of integration, and testing efforts.

Point to point ensures that one application would consume the message at a time. For example, it is used in RPC to transfer data.

Pattern – Rich Integration Interaction Model

Service and semantics standardization is mainly focused on abstracting the complexities by targeting flexible and loosely coupled connections.

Composite service is focused on building flexibility into the services themselves in each of the dimensions of policy, process, and structure by hiding the multiple fine-grained services exposed by legacy systems and creating business-aligned coarse-grained services.

Process Service is similar to composite service but with larger scope. Its focus is on hosting end-to-end business processes involving systems and people, which represents an end-to-end business capability delivered by business.

Pattern – Multichannel Application Interaction Model

Multichannel services provide faster response to business change requiring the support of new channels; consistency of capability and information is delivered to all channels.

Multichannel and differentiated services provide faster response to business change requiring the support of new channels; consistency of capability and information is delivered to all channels. In addition, they provide added value to the privileged consumer while minimizing the price. The focus is on consistency across divergent interaction methods and technologies, building a generally related body of services and addressing specific issues of security and versioning.

Pattern – External Partner Integration Interaction Model

The focus is predominantly on security, closely followed by protocol translation, QoS, and semantic adaptation. Third-party applications are integrated with the DXP's application.

Pattern – Event-Driven Adaptive Enterprise

Handling opportunity, disruption, and threat: The focus is to capture events from multiple, loosely coupled systems and analyze them based on the situational, temporal context such as time-series analysis of system threats, system logs, etc. and respond based on the predefined rules.

End-to-end process visibility with KPI monitoring: The focus is to capture events from multiple loosely coupled systems and analyze it based on the situational, temporal context such as time-series analysis of their key business metric and create visibility into key business processes.

Data Standards

Data standardization would help you to develop and agree on the most appropriate standards for the business requirement from various protocols and architecture, which further constitutes different payload types and data structures or formats. You can use a serialization mechanism to handle inconsistency of data structures and their varying payloads, which would translate data from one data structure to another and deserialize it, that is, extracting the data structure from data itself. Serialization is the process of converting an object into a stream of bytes to store the object or transmit it to memory, a database, or a file. It saves the state of an object so that it will be able to recreate it. The reverse process is called deserialization.

The most commonly used protocol and architectural pattern are SOAP and REST, respectively. SOAP is a protocol, which has a WSDL file and contains information about web services and the location of the server. It uses service interface to expose its functionality and it requires more bandwidth. To make data communication and transformation more flexible and light, REST was introduced, which contains features like stateless, cacheable, layered, uniform interface. REST is lightweight and mostly contains JSON messages, therefore the size of the message is much smaller than with SOAP. The DXP integration layer consumes the data in any format but provides RESTFUL-JSON services to the DXP UI layer, which ensures lightweight and flexible data interoperability.

Flexible Integration Middleware

The DXP concept is to support different middleware integration frameworks, which makes the DXP integration layer flexible so that the data and message are exchanged with multiple systems, irrespective of the technology and framework used in existing systems.

Delving into the methodological world of services and integration of services along with to fulfill the integration requirement of EAI (enterprise application integration), ESB was introduced. It was considered as a central hub for integrating services but it is a single point of failure. So, to eliminate this issue, SOA (service oriented architecture) was introduced that, further enhanced with an ESB, has the flexible and distributed capabilities to solve the integration problems by implementing EIP (Enterprise Integration Patterns).

Nowadays we are moving toward the microservices that build on lean structure, where services are developed, deployed, scaled, and maintained independently, which further enhances the business requirement, and time to production or market is reduced significantly.

EAI vs. SOA vs. ESB vs. Microservices

ESB and microservices are models based on SOA. The service-oriented model is an implementation to achieve enterprise application integration.

- EAI is a framework that connects and integrates the different application and data source in an organization to simplify the business process.

- The EAI framework provides cross-platform and cross-language integration to simplify the business service by exchanging information between the different applications.

- SOA is an integration paradigm that is based on design principle of architectural interoperable services, which deals with data sources, software, and message processing.

- ESB is a software architecture that provides integration of enterprise applications and services for complex architectures.

CHAPTER 5   DESIGNING THE INTEGRATION LAYER

- ESB focuses on interaction and communication between components, to handle data transfer or message exchange between services. ESB is capable of transforming message data into a format that the application understands.

In Figure 5-2, multiple applications are communicating and exchanging data in different formats using broker message communication.

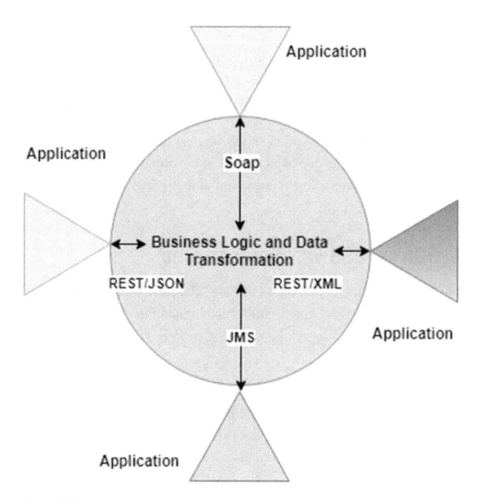

Figure 5-2. *ESB*

Key requirements to fulfill the integration layer objectives are as follows:

Mutual Memorandum of Understanding (MOU)

Service contracts or an MOU are useful before developing integration services, so as to ensure QoS, security, and scalability of services. Initially, the MOU was defined with a SOAP interface. Several frameworks and tools supported SOAP, but nowadays after the increasing popularity of Restful services, Swagger is becoming the most vital standard for defining, implementing, discovering, and testing REST services. Swagger-enabled API provides interactive documentation and software developer's kit (SDK) generation capabilities.

Service Protocol and Data Format

While developing services, you should check the service communication endpoints, for example, HTTP endpoints for HTTP protocol-based applications; JMS endpoints for JMS applications; machine-to-machine (M2M) data transfer protocol (MQTT) endpoints for MQTT-based applications; and data format, for example, XML and JSON requirements.

API Management

API gateways are used to manage API. You have to consider the requirement, whether services are developed to use within the application or would be used by multiple applications across an organization.

Why Do We Need Data Transformation Capabilities in DXP?

In a business organization there are multiple applications interacting with each other, therefore the data will be organized and structured in different forms. For example, old systems are built on SOAP architecture but a DXP consumes REST services; hence, to make an old system compatible with a new modernized system, we need data transformation capabilities, whereas the DXP integration layer is capable of integrating with other systems irrespective of technology and framework. The DXP integration layer is capable of consuming data in any format and converts the data as per the DXP UI requirements, that is, data can be in any format. For example, SOAP, XML, JSON, REST, comma-separated values, MS-Excels, databases, etc. would be transformed as per the DXP UI requirement, for example, Restful-JSON. However, without changing existing systems we can consume the services from the existing system and transform the data as per the DXP UI requirement, which makes the DXP integration lean and flexible.

Integration Technology Stack and Architecture

The integration technology stack depends upon integration architectural patterns, which are monolithic and microservices architecture. You would choose for your application while developing the DXP's integration layer.

Monolithic

A monolithic architectural integration technology stack contains one of the integration frameworks according to your integration requirements, such as Apache Camel or other integration framework, as listed in Figure 5-3. These integration frameworks consume the data API from other third-party applications, transform the data according to the client application, and expose the API to the client application. Services such as mobile services, portal services, OTP services, ERP services, master data management (MDM) services, etc. are part of one single application.

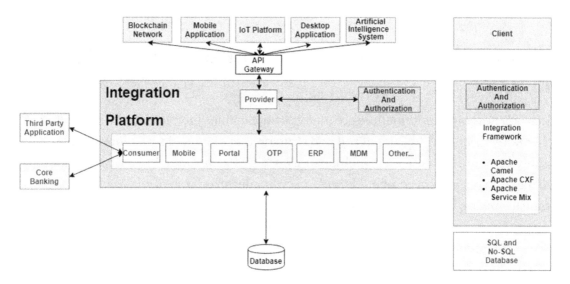

Figure 5-3. Monolithic integration

CHAPTER 5 DESIGNING THE INTEGRATION LAYER

The API gateway exposes the services to the client application; the authentication and authorization module is part of the integration platform.

The integration layer consists of API providers and API consumers. The API consumer consumes data and messages from other applications and implements authentication and authorization using respective modules, along with the integration logic and business logic related to particular services as per the API provider, which is consumed by the DXP UI layer (mobile and desktop client applications), other applications and platforms such as a blockchain network, IoT platforms, and AI platforms.

Multiple integration services like social media services, analytics services, OTP services, and third-party services are consumed by the API consumer, whereas the API provider provides transformed API to the DXP client application. As shown in Figure 5-4, API consumers are responsible for consuming the data and consumed data is transformed by the integration layer, and API providers are responsible for exposing services to the UI layer.

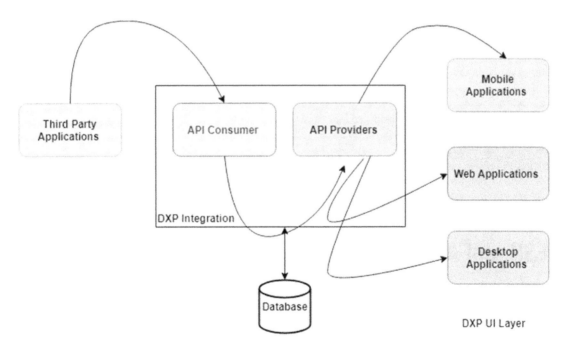

Figure 5-4. *DXP monolithic integration*

CHAPTER 5 DESIGNING THE INTEGRATION LAYER

Microservices

The microservices architectural integration technology stack contains a suite of small services, each running its own process and communicating with lightweight mechanisms on an individual port number. One service would contain one business capability, as shown in Figure 5-5. OTP services have OTP capabilities; similarly, other services have their own process and lifecycle. These services are separately deployable, hence a faster release cycle. You can use different microservices frameworks such as SpringBoot, Lagom, etc., as shown in Figure 5-5.

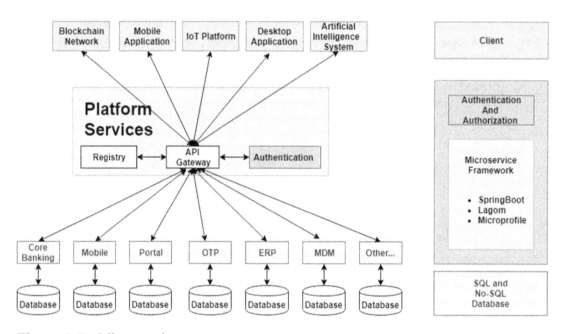

Figure 5-5. Microservices

ESB and API Gateway

An ESB would be used for integration, orchestration of multiple services into one service, routing of services, and event handling and monitoring. An ESB is based on service-oriented architecture, which is an efficient service delivery platform. On top of an ESB, you could use a service gateway for security, policy enforcement, and exposing services as an open API to external consumers (public). A service gateway manages your integration services built with an ESB.

It is recommended that web services need to be pushed via an API gateway. The API gateway ensures security, as it uses open standard authentication frameworks and protocols such as SAML, Kerberos, Oauth, etc.

Integration Security

Security is the main concern when exposing an API to a client application over web. There are multiple frameworks and protocols that provide different aspects of security such as authentication, authorization, and integrity.

Authentication and Authorization

You can use different frameworks and protocols to authenticate and authorize the user on the basis of tokens and session to access the application, as per the requirements. We will become familiar with these frameworks and protocols in the following subsections.

Protocols

Authentication protocols are a type of cryptography protocols designed for transfer of authentication data between two entities in a secure way. Single-sign-on (SSO) can be achieved using these frameworks, where the user presents information (user ID and password) once and gets an access token that is valid to access all connected applications in the environment for a particular session. For example, Kerberos, NTLM, OpenID, and SAML are the most common protocols, which provide features like SSO by providing an access token that is valid for all the applications integrated with these protocol. This access token contains the authentication and authorization information of the user for a particular session.

Frameworks

You can use authentication and authorization frameworks like Oauth, Spring Security, etc. These frameworks provide a mechanism to integrate existing protocols and provide security implementation to your application. You can also use two-factor authentication (2FA) frameworks such as Google Authenticator, one-time password (OTP) authentication on mobile, Duo, Authy 2FA, etc. with other security frameworks to provide an additional level of security to the user.

CHAPTER 5 DESIGNING THE INTEGRATION LAYER

Session Management

Session management is a way to manage the state of the application. A DXP uses HTTP protocol to provide data and persistence services to applications because HTTP is a stateless protocol; stateless means that the server can send client requests to any node in the clusters while load balancing the application. Each time, a user's request is independent because there is no state; a user request can be distributed to any server, so the way to maintain the state of the application between client and server is to use the session on the server side to save the user state because the server is stateful. There are different mechanisms to maintain the state of the application, such as session stickiness, session replication, and shared or centralized session. We can use sticky session to ensure that all requests from the specific user are sent to the particular server through a load balancer. But in case that particular server goes down, and the load balancer is forced suddenly to shift the user to a different server, all of the user's session data would be lost. To overcome this problem, a session replication mechanism can be used;: that means each instance saves all the session data and synchronizes through the network using a library such as Jgroups, Hazelcast, Redis, etc. but synchronizing session data causes network bandwidth overhead. To overcome this problem, you can use a centralized session storage mechanism; that means whenever a user accesses any services, user data can be obtained from shared session storage. In some use cases this scheme works excellently and can be achieved using the JDBC (database) session storage mechanism so that all servers can access the same session object stored in the database.

Token Management

It is recommended to use session management at the server because the servers' applications are stateful, and tokens management at the client to store user login status. Tokens are held by the users themselves and are stored in the browser cache or in the form of cookies. Each time a request is sent to the server, the server can check the identity of the user and determine whether it has access to the requested resource. As the token is used to determine identity, the content of the token needs to be encrypted to avoid security attacks; this can be achieved using standards like Java Web Token (JWT) which is open-standard (RFC 7519) and defines the token format and contents. It can encrypt content using various asymmetric and symmetric encryptions as per requirement. This ensures the integrity of data transferred between two parties, that is, client application and server application.

Integration Best Practices

You should follow best practices while developing the integration layer as mentioned in Table 5-2.

Table 5-2. *Best Practices*

	Area	Best practice
1	Service design	• Services must be designed to behave as stateless services, since there is no grid technology at this point. • Services must not persist any session-related information or transient state of the request in memory, that is, use of shared variables or local cache must be avoided completely. • Services must be designed to handle duplicates. If for some reason the same request message is transmitted twice, the service must enforce the messaging semantics in order to identify the duplicates and reject them. • It is possible that there could be multiple instances of the service operation running concurrently. Service must share resources (file handles, socket connections, etc.) in a thread-safe manner, avoiding deadlocks. • If an exception occurs in a subprocess, then the typical practice is to propagate that exception to the parent process either through a "Generate Error" activity or "rethrow" exception. But, if the sub process is nested deep below from the main process, this could cause a problem since every subprocess must rethrow the error. It is a known fact that rethrow of exceptions is costly, since the entire copy of the stack must be embedded with each throw. Hence, it is recommended that we report these exceptions by setting flags like "Exception=true" and exiting the process with proper error handling. This should be propagated to the parent process in an optimal manner.

(continued)

Table 5-2. (*continued*)

	Area	Best practice

2	Process and activity design	• Avoid extensive "call process" hopping, or a long chain of process calls for a single end-to-end integration.
		• Prefer process starters over wait-for activities and allow for parallel processing. Process starter fits better for BusinessWorks (BW's) threading model because flow control is not available for wait-for activities.
		• Reduce the number of activities, if possible.
		  • Each task requires overhead and reduces performance.
		  • For example: If data could be mapped in a SOAP request activity, do not use a mapper activity only to map data for the SOAP request.
		• Database operation
		  • Database operation in a nontransaction group
		    • Set maximum db connections = engine thread count value and set in Admin.
		  • Database operations in a transaction group
		    • Set maximum db connections = total number of expected concurrent transaction groups.
		  • These parameters should be controlled through the flow control properties.
		  • Use Batching instead of Statement when possible to improve latency.
		  • Indexing is a must.

3	Global variables	• Global variables (GVs) are kept in the memory in an XML structure and user does not need to worry about these getting stale. • Remove unused GVs periodically. • GVs are read-only, hence could be accessed concurrently. No synchronized access required. • Accessing the last element takes longer than the first element. So, remove unused GVs and arrange the rest from most frequently used (MFU) to least frequently used (LFU).
4	Transport options	• HTTP is better performing than JMS on BW; SOAP protocol adds significant overhead, whether the transport is HTTP or JMS. • SOAP over HTTP has better throughput than SOAP over JMS.

(*continued*)

CHAPTER 5 DESIGNING THE INTEGRATION LAYER

Table 5-2. (*continued*)

	Area	Best practice
5	Version control of BW process	As a best practice, use the version controller of the BW processes during the software development life cycle (SDLC); version control systems should be configured. Designer project should be checked-in to enterprise version of the configuration management system.
6	Template-based development	ESB designer supports template-based development of the BW processes. Hence, a common template should be defined for services and processes, which are similar in nature.
		This template should be the baseline and all the processes should be developed using this process. Usage of the template not only reduces development time but enforces common standards of development.
7	BusinessWorks naming standards	The structure is: <BusinessDomain><BusinessSub-domain><ServiceName> • Business domain of the service in the enterprise • Business subdomain of the service in the enterprise • The name of the service

8	Component naming standards	Process naming standard
		 • Process Starters - Receive<object><event>
		 • Service - <ServiceName>
		 • Process - Operation Name - <ServiceName><OperationName>
		 • Subprocesses - <action><object>
		   • Database persisting: Persist<object>
		   • Converting XML to a string: Render<object>
		   • Sending to BC: Send<object>
		   • Transform: Xfrm<object>To<object>
		 • Validation processes - Validate: Validate<object>

9	Enterprise archive naming standards	• <ProjectName>_<DeploymentClassifier>.ear
		An enterprise archive (EAR) file would be created for each business process engine to be deployed.
		• <ProjectName>_<DeploymentClassifier>.war
		A Web application ARchive (WAR) file would be created for each business process engine to be deployed as a web application.

CHAPTER 5   DESIGNING THE INTEGRATION LAYER

BXP Case Study

BXP integration concepts enable the organization to integrate the data from a third-party application, so that different users logged in to a single portal could access their details, transactions, and accounts. The banking domain contains different third-party applications (master data management, core banking application, OTP and SMS gateway, third-party authorization application through LDAP server, etc.) to run the whole banking organization, as shown in Figure 5-6.

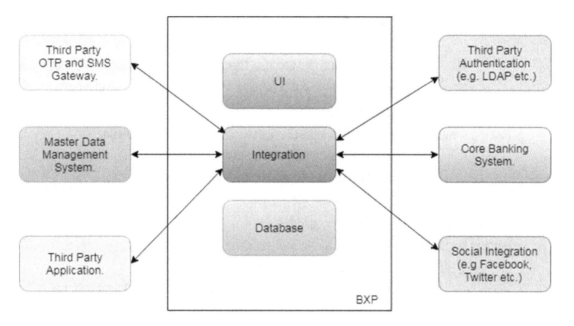

Figure 5-6. *BXP overview*

Integration services of a BXP integration layer have brokers, which communicate between BXP UI components and third-party services and systems. A BXP implements the well-known EIP and therefore offers a domain-specific language, standardized to integrate applications. As shown in Figure 5-7, account and transaction services are getting transformed. Account is the service exposed by the MDM system in JSON format over HTTP, whereas Transaction is the services exposed by the core banking system in XML format over HTTP. But the DXP UI application needs data in JSON format over HTTP protocol, hence the integration layer is introduced to handle data transformation and expose the new API as getAccountDeatils and getTransactionDetails to the UI application.

- *API consumers*: API consumers consume data from different applications that consists of different data formats and payloads in the form of message exchange. The integration framework and its capabilities help to transform the data according to business requirement using EIP.

- *API provider*: The API provider provides newly transformed services exposed with new endpoint for DXP UI components. The API provider exposes the REST endpoint in JSON format as per the DXP UI requirement.

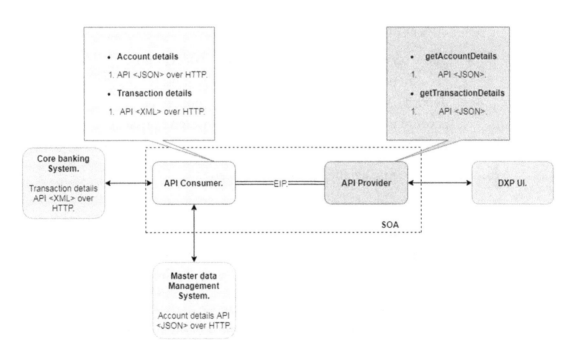

Figure 5-7. Data transformation

We will look into this integration problem in brief. A banking organization has two services, that is, account details and transaction details, available from two different systems in different formats. But the BXP needs integration so that data would be populated on BXP UI components as per BXP requirements. To transform these services, let's look into the BXP UI requirement.

- *BXP UI requirement.* A list of BXP-UI requirements follows:

    a. Data should be in JSON format.

    b. UI components need limited data points: customer name, account number, current balance, and last ten transactions.

- Existing services. A list of services exposed by the existing system follows:

    a. Core banking application (transaction services)

    - Data available in JSON format

    - These services contain a huge number of data points: account number, last one month's transactions, FD (fixed deposit) details, etc. But the DXP UI component requires account number, current balance, and last ten transactions for a particular user.

    b. Master data management (MDM) application

    - Data is available in XML format.

    - Huge number of data points: customer name, customer number, service number, address, etc. But the DXP UI component requires customer name and account number held by that customer.

The integration layer consumes required fields from aforementioned existing services, transforms the data format using EIP patterns, and exposes getTransactionDetails and getAccountDetails as new REST services with required and limited data points.

Sample code on developing integration layer is available in Appendix B of the book. We have chosen Apache Camel framework as the DXP Integration Layer because it supports a large variety of EIP patterns to transform the data; it is open source, scalable, and flexible to support DXP concepts.

Case Study Conclusion

You should be able to understand the different integration services and integration of these services to a DXP using EIP. Underscoring the various protocols, architectural pattern, and data structure should help you to design an integration system using DXP concepts and principles. The BXP case study should help you to understand the design principles of the DXP integration layer where one has modeled two services and exposed these web services as REST API to BXP UI components.

Chapter Summary

- The DXP integration layer is the most important part of a DXP. Integration is the backbone of the enterprise application, which provides an efficient and flexible integration layer to provide efficient data interpretability.

- You understand the DXP integration principles and core concepts.

- You will be able to analyze business requirements to develop an integration layer as per DXP core concepts.

PART III

Securing the Banking Experience Platform

CHAPTER 6

DXP Security

Digital experience platforms (DXPs) consist of various technologies such as web technologies, content technologies, and database technologies. Security is one of the fundamental attributes for the success of a DXP program. Lack of adequate security measures impacts the users' trust and drastically impacts the application. Defining and maintaining DXP security is a continuous process.

In this chapter we discuss the core security concerns of DXP.

DXP Security Framework

The DXP security framework defines the key tenets of security concerns that need to be addressed in a DXP application (Figure 6-1). We need to address all categories belonging to each of the security tenets.

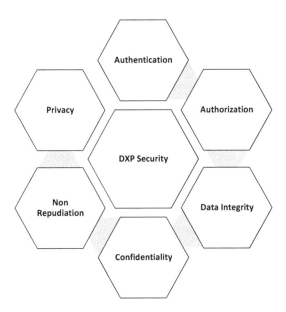

Figure 6-1. DXP security framework

The key elements of the DXP security framework are as follows:

- Authentication validates the user's identity. This includes the authentication mechanism, single-sign-on (SSO), and management of user credentials.

- Privacy management ensures that the user's personal information is protected. Privacy concerns should be addressed while storing the information, during transit, and sharing the information with third-party services.

- Authorization validates the user's permission and fine-grained access privileges to functionality and resources. Authorization enforces the privilege/role-based access to secure resources.

- Confidentiality ensures that information exchange between intended parties is done securely on a need to know basis.

  - Data integrity ensures that information is not modified during transmission. Encryption and secure transport are needed to guarantee the data integrity.

  - Nonrepudiation ensures that data and proof cannot be altered or deleted, using robust tracing, authentication and authorization processes, and auditing.

DXP Layer-Wise Security

As a DXP is built with various layers, it is necessary to enforce security at each of the layers.

Layer-wise security vulnerabilities and security measures for the DXP are shown in Figure 6-2.

CHAPTER 6 DXP SECURITY

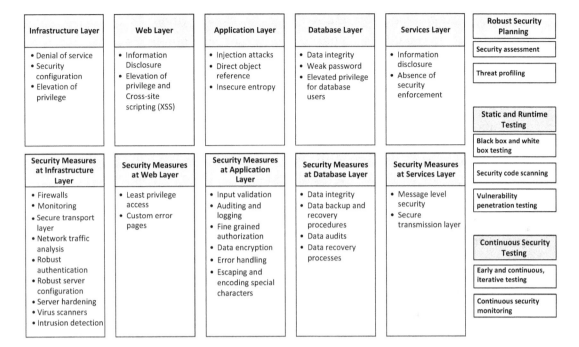

Figure 6-2. *DXP layer-wise security vulnerabilities and security measures*

The following list details various security measures and best practices that can be taken at each of the layers:

- *Infrastructure layer security*: The main security vulnerabilities at this layer are denial of service attack, incorrect security configuration, and elevation of privilege. In order to address the security vulnerabilities, the appropriate security measures include firewalls, robust monitoring, and usage of SSL (Secure Sockets Layer)/TLS (Transport Layer Security), providing robust server configuration, server hardening (disabling of all unnecessary ports, services, and protocols at the server machines), and installing virus scanners and intrusion detection systems.

- *Web server layer security*: The main vulnerabilities at this layer are accidental information disclosure in the stacktraces and error messages, and elevation of privilege. In order to address this, we need to use custom messages that hide the system and framework details. Enforce a least privilege policy for all resources to prevent accidental information leakage, minimizing the risk of escalation of privileges.

- *Application server layer security*: The main security vulnerabilities at this layer are injection attacks, and insecure encryption methods. The security best practices to mitigate the security threats at this layer are robust input validation; enforcing right access to resources; proper error handling; robust encryption methods; and using robust auditing, monitoring, and logging.

- *Database server layer security*: The main security vulnerabilities at this layer are weak password and elevation of privilege. In order to address this, we need to enforce least privilege access to the application database user and establish robust data backup and recovery processes.

- *Services server layer security*: At this layer the main security concerns are information disclosure and absence of encryption measures. In order to address this, we need to enforce message level security such as encrypting messages, adding security tokens (encrypted information consisting of details such as logged-in user role, timestamp, and such) to the messages, and using a secure transport layer for message communication.

In addition to these layer-wise security measures, we should also include the following in security best practices:

- *Robust security planning*: We need to do a detailed security requirement assessment of the DXP application and do the threat profiling. Threat profiling will help us to identify the security scenarios and we can develop test cases based on that.

- *Static and runtime security testing*: Static security testing includes automated security testing and secure code review; runtime security testing includes penetration testing.

- *Continuous security testing*: Security testing should be carried out throughout the project lifecycle on an iterative basis. A real-time security monitoring infrastructure should be set up to continuously monitor security incidents.

Common Security Scenarios of DXP

A DXP is mainly built on web technologies. Hence all the threat scenarios applicable for the Web are also relevant for DXP.

We have listed the common security best practices that can be used in DXP implementations as follows:

Password Standards

The password policies of the DXP should enforce stricter password rules during account setup, password change, and all account verification scenarios. Insufficient or weak password policies lead to increased vulnerability. The key best practices for password policies are given below:

- Enforcing minimum password length. Normally a minimum password length of 8 is suggested.

- Enforcing the mix of numeric values, special characters, and uppercase letters in password text

- Forcing password change regularly (for instance forcing user to change password after 90 days)

- Avoiding common passwords or dictionary terms or common phrases as passwords

- Maintaining a history of previous passwords to ensure that new password does not repeat from the history

- E-mail password reset link instead of mailing the updated password in plain text.

- Password should be stored as one-way hash (the ones that cannot be decrypted) while storing in a database or properties file. Use strong encryption algorithms such as AES 128 or SHA1 256 bit encryption mode.

- Audit password retries and restrict the maximum number of password retries.

- Use CAPTCHA for functions such as user registration and password reset, to prevent automated attacks and bot-based attacks.

- Lock the account after a specified number of successive failed password attempts.

Session Management

A session is established once the user successfully logs in. A valid user session consists of user information, application data, and such. The main best practices in managing user session are given as follows:

- Enforce automatic idle session timeout that invalidates the session after a specific duration of inactivity. This prevents accidental misuse of a session. Usually a 30-minute window is suitable for idle session timeout.

- Prevent multiple sessions/user logins at the same period to prevent session hacking.

- Use CSRF (cross-site request forgery) tokens (a unique ID) along with each request to prevent CSRF attacks.

- Do not store any sensitive information in session cookies.

- Use a secure transport protocol (such as HTTPS, FTPS) while transmitting sensitive information.

- Don't send the session IDs as URL parameters.

Information Management

The information related to a DXP system and the application data should be carefully guarded to prevent sophisticated hacking attempts. The following are the key best practices related to information management:

- Prevent accidental disclosure of the information in log files, exception messages, and error logs. This includes information related to web server name/version, programming language used, host name, IP address, and such sensitive information.

- Develop a data classification policy based on the sensitive nature of the data. For instance, we can classify the data into three categories: "public," "private," and "confidential." We can then apply various security policies based on the category.

- All sensitive data such as passwords and server names should be encrypted before storing or during transmission.

- Update all server configuration to hide the server-specific information and create a custom error page.

Data Validation

DXP-based applications receive data from the end user in many scenarios, such as registration data, review comments, and such. All such end user data should be properly validated to prevent various attacks. The following are some of the best practices of input data validation:

- Validate all user input data, using white list and black list values. A whitelist provides a list of all allowed characters and a blacklist provides a list of all disallowed characters. Encode or escape reserved and special characters (such as HTML tags or JavaScript code).

- Validate end user values while using them for executing database queries. This prevents SQL injection attacks.

- Perform strict validation on input data received from end users and nontrusted sources. This includes length validation, special character validation, blacklist validation, type valuation, format validation, range validation, and others.

Service Security Management

Services are widely used in DXPs. The core integration layer in a DXP is built around services. Services-based architecture provides loosely coupled layers that can be easily extended. The main best practices in services security are given as follows:

- Use a secure transport layer such as HTTPS to provide transport level security. Secure transport layers use certificates that ensure data integrity and prevent any data interception attacks.

- The message level encryption includes encrypting the message or signing the message with a digital signature to ensure message confidentiality.

- Add security headers to the message envelope and/or add security tokens to the message.

- Use open security standards such as ws-security or SAML (Security Assertion Markup Language) to enforce services security.

Security Vulnerabilities and Best Practices of DXP

Table 6-1 provides the key vulnerabilities and the main best practices to address the vulnerabilities.

Table 6-1. *DXP Security Vulnerabilities*

Category	Main Vulnerabilities	Best Practices
Session management	• Predictable session IDs • Absence of session timeout • Man in middle attack • Cross-site request forgery (CSRF) • Denial of service (DoS) and distributed denial of service (DDoS)	• Prevent multiple simultaneous logins. • Ensure session ID randomness through strong random number generators. • Provide auto session timeout. • Use secure transport layer (SSL/TLS) for confidential information. • Don't store sensitive information in cookies. • Use secure attribute of cookies. • Don't use HTTP GET requests for sensitive data. • Don't provide administrative interfaces to Internet users without strong authorization controls. • Only enable the needed HTTP methods. Methods such as TRACE, PUT, and HEAD can be disabled if not used by the application. • Enable WebDAV methods only for the authorized content authors.

(*continued*)

Table 6-1. (*continued*)

Category	Main Vulnerabilities	Best Practices
Data validation	• Injection attacks (related to SQL, LDAP, OS, XPath, HTTP header) • Cross-site scripting (XSS) • Executable file uploads	• Perform client-side and server-side validation • Encode and sanitize end user data. • Validate against blacklist and reject/remove all special characters. • Use parameterized queries or prepared statements or ORM frameworks for database query execution. • Block any scripts calling OS commands and prevent upload of any executables. • Use XSS filters to allow only safe commands and sanitize inputs.
Authentication	• Brute force attack • Weak password policy • CSRF	• Usage of CAPTCHA • Enforcing strict password policies • Use CSRF tokens with web requests. You could use referrer validation or custom HTTP header (such as X-requested-by) to avoid CSRF attacks. • Use multifactor authentication (MFA) for secure operations. • Disable autocomplete for form fields storing sensitive information (such as user names, passwords, email IDs, etc.)
Authorization	• Absence of sufficient authorization and permission model • Insecure user impersonation	• Define role-based access and permission model. • Validate user details before sensitive operations such as password update. • Use "deny by default" and "least privilege" policy. • Implement fine-grained access control to resources such as pages and functionality.

(*continued*)

Table 6-1. (*continued*)

Category	Main Vulnerabilities	Best Practices
Information management	• Accidental information disclosure	• Remove all server-specific information from log files. • Display only minimal error information during security incidents. • Only collect required personally identifiable information (PII) and secure PII at rest and during transit. • Don't cache sensitive data. • Use defense in depth principle performing access and data validation at all layers.
Infrastructure security	• DoS • DDoS • Elevation of privilege	• Harden the servers and only allow traffic on preconfigured ports. • Run the application process with only necessary and nonadmin privilege. • Provide least privilege for service login accounts that run application processes. • Use web application firewall (WAF) to prevent web-based attacks. • Disable all unnecessary services and install an antivirus scanner on server machines. • Use load balancer and/or cloud security to handle the surge in traffic. • Filter packets at the firewall to prevent DoS/DDoS attacks.

Security Testing Framework for DXP

Security test planning for DXP should be done iteratively in all phases. For comprehensive security we should continuously monitor the application security events in real time and take corrective actions during security incidents.

The key steps of a DXP security testing framework are given in Figure 6-3.

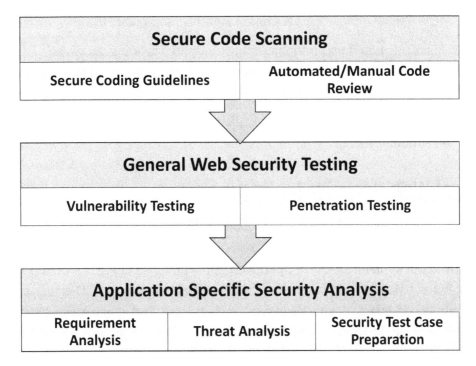

Figure 6-3. DXP security testing framework

The main stages of DXP security testing are as follows:

Secure Code Scanning

In this stage the development team uses secure coding guidelines and standards to develop the DXP application. Security guidelines and coding standards and white box security tools are used in this stage. Automated security code scanners and manual code reviews can be used to check for any known security issues.

The common security issues that can be uncovered through security code review are listed as follows:

- Usage of password in plain text format in code or in configuration file or in database

- Use of the user inputs directly for SQL queries or LDAP queries, leading to injection attacks

- Absence of user input validation, cookie values, URL parameters, and form field values

- Use of weak unique key generation algorithms to create IDs

- Stack traces and exception handling modules providing details of server and other internal details that can be exploited by others

- Absence of auditing and monitoring security events such as account lockout, login, logout, password failed attempts, password change events, and such.

General Web Security testing

In this stage, we do the black box security testing for common and known vulnerabilities. We carry out testing of vulnerability scenarios and penetration scenarios. This includes testing the application for OWASP (Open Web Application Security Project) top 10 vulnerabilities, CWS/SANS (SysAdmin, Audit, Network, Security) top 25 errors, and common injection attacks. Many tools such as Burp Suite, Zed attack proxy, Fiddler, and WebScarab can be used for vulnerability testing and penetration testing. The main testing scenarios in this category are as follows:

- XSS testing

- Testing of directory browsing of resources

- Testing of absence of access controls on protected resources.

- Testing of access to URLs and resources

- Checking for accidental information disclosure in cookies and HTTP headers

- Checking for information leakage in server error pages and error handling

- Checking of CSRF tokens

- Checking for misconfigured security settings such as misconfigured HTTP headers or misconfigured error pages

- Testing for buffer overflow

- Testing the injection attacks (SQL injection, LDAP injection, XPath injection)

- Testing for denial of service

Application-Specific Security Analysis

In this stage we analyze the requirements of the DXP application and understand the security-related requirements. We perform the threat modeling by understanding the details of the following:

- Identifying the main security objectives of the organization and the application
- Types of users and their security needs
- Data security needs
- Details of sensitive operations
- Details of client-side and server-side validations
- Details of authentication and authorization

Based on this, we will create security test cases. Let's look at the threat profile of two key operations in a banking domain. We have considered "transaction management" and "funds transfer" scenarios for threat profiling here.

Threat Profiling of Transaction Management in Banking DXP

The main vulnerabilities in transaction management are as follows. We can create security test cases for testing these scenarios:

- Using SQL injection attacks to tamper the transaction details (account number, account holder name, timestamp) and attempt to view/update/delete the transaction details of other users
- Adding dummy or duplicate or incomplete or inaccurate transactions
- Attempting to steal transaction details of other users through man-in-middle or CSRF attacks

CHAPTER 6　DXP SECURITY

Threat profiling of Fund Management in Banking DXP

The main vulnerabilities in fund management are as follows. We can create security test cases for testing these scenarios:

- Altering the funds balance in the account through SQL injection attacks
- Attempting to transfer funds from other users account
- Changing profile details of other users
- Viewing account details of other users

DXP Security Checklists

In this section we have defined the checklist for various categories. The checklist can be used during security code review and during security testing.

DXP Architecture and Design Phases Security Checklist

- Ensure that security validations are done at client-side, server-side, and all integration layers.
- While using the third-party party libraries, check for any known security vulnerabilities.
- Identify all the sensitive data and categorize the data based on their sensitiveness. Clearly define the security policies for each of the categories.
- Enforce continuous and iterative security testing at all SDLC lifecycle stages.
- Define a robust cryptography process and encryption process. Prefer strong encryption algorithms such as SHA 256 bit for encryption and use one-way hash algorithms for encrypting sensitive data.
- Establish a secure transport layer for all sensitive transactions.
- Perform detailed thread profiling for the DXP application and define the security test cases for the same.

DXP Information Management Security Checklist

- Define the security categories for the DXP application. For instance, we could define three categories such as public, private, and confidential. Define security policies for each of the security categories: all data marked as public is visible to guest users and public users; all data marked as private is accessible only to logged-in users; and all data categorized as "confidential" should be visible for users with admin role. Also, all data categorized as "private" or "confidential" should be encrypted during rest and during transit.

- Ensure that no sensitive data (such as SSN number, user passwords, credit cards, etc.) is stored application logs in plain text.

- Ensure that no sensitive data is cached or stored in browser cookies. Disable browser autocomplete for sensitive form fields.

- Ensure that no sensitive data is shared with external or third-party party services without consent from respective owners.

- Don't use sensitive data in hidden form fields, meta tags, or custom HTTP headers that are accessible to the end user.

DXP Authentication and Session Management Checklist

- Ensure that there is a centralized authentication management system that uses enterprise-wide LDAP or Active Directory.

- Ensure that the centralized authentication system enforces robust password policies. Password policies include restriction on password strength, password change frequency, notification on password change events, account lockout policy, and such.

- Define multi factor authentication (MFA) for sensitive functions such as admin functionality, password update functionality, and such.

- Define strict session management policies including idle session timeout, avoiding multiple simultaneous sessions, creating random session IDs, and such.

- The DXP application processes must run with minimum privilege.

- Invalidate the session upon logout. For SSO scenarios, invalidate all sessions across logged-in enterprise applications.

- Use secure cookie attributes such as HTTPOnly and secure and strict-transport-security header, to ensure secure data transmission.

- Use the principle of least privilege for accessing secured resources.

- Use CAPTCHA to minimize automated form submissions, bot based attacks, and password resets.

DXP Network Communication Management Security Checklist

- Establish end-to-end TLS and use SSL communication for all sensitive data communications.

- Leverage firewalls and configure rules to block the packets and traffic that lead to denial of service attacks.

- Establish a network monitoring infrastructure to identify any security incidents in real time.

- Don't allow unsecured protocols (such as HTTP) to sensitive resources such as sensitive web pages, URLs, data, and functions.

- Use only valid SSL certificates signed from an authorized certification authority (CA).

DXP Input Validation Security Checklist

- Sanitize the user input data and remove and encode all special characters that lead to injection attacks and XSS attacks. This includes submitted form data, user-generated content (UGC) such as blog posts, review comments and such, and URL parameters.

- Encode the HTML response especially for the UGC.

- Use unique CSRF token with each request to prevent a CSRF attack.

- Prevent upload of executable files and validate all the uploaded files.

DXP Security Auditing and Logging Security Checklist

- Ensure that the application logs all security events such as password changes, login failures, administrative activities, role/permission changes, and such.

- Don't allow the default framework error messages and stack traces (log messages) to appear in the error pages. Display generic error messages that don't reveal details about the application framework and technologies.

- Do not log the sensitive data in the application log files.

- Secure the logs files and restrict the access to the log file location.

Chapter Summary

- The DXP security framework defines the key tenets of security concerns that need to be addressed in a DXP application.

- The key elements of the DXP security framework are authentication, authorization, privacy, integrity, nonrepudiation, and confidentiality.

- Authentication validates the user's identity.

- Privacy ensures that the user's personal information is protected.

- Authorization provides role-based access to functionality and resources.

- Confidentiality ensures access is granted only to privileged users.

- Integrity ensures that information is not modified during transmission.

- Nonrepudiation ensures that that evidence cannot be altered or deleted.

- Layer-wise DXP security includes enforcing security best practices at the infrastructure layer, web server layer, application server layer, database layer, and services layer.

- Common security scenarios of a DXP include password standards, session management, information management, data validation, and service security management.

- The key steps of the DXP security testing framework include secure code scanning, general web security testing, and application-specific security analysis.

CHAPTER 7

DXP Information Security

Information security is crucial for digital platforms that are used for financial domains such as banking. In this chapter we discuss various aspects of information security. The best practices given in this chapter can be used for defining and implementing a robust information security framework for DXPs.

Information Security in DXP Solutions

Information security defines policies for protecting data at rest and data during transit. The basic principles of information security are defined as follows:

- *Information security policies*: Organizations should define security policies and procedures and processes to protect information from unauthorized access and appropriate use.

- *Data access policy definition*: Access policies should be defined for data, based on the sensitive nature of the data.

- *Defense in depth*: Provide layer-wise access policies at each of the application tiers.

- *Compartmentalization*: Group the information and provide access to grouped information (called "compartments") only on a need-to-know basis. Compartmentalization reduces the attack surface and can be implemented using layer-wise security and the least privilege principle.

- *Least privilege by default*: Provide only the minimal needed privileges for entities and processes.

- *Centralized access*: All security policies such as authentication and authorization should be centrally controlled.

Implementing Defense in Depth

Defense in depth provides security checks at each of the layers. We will explore defense in depth at all layers.

Firewalls and Proxies

For untrusted zones and external facing systems, we need to use a firewall that provides protections against attacks, as listed in the following:

- Denial of service (DoS) and distributed denial of service (DDoS)
- Protection against spam and malware
- Load balancing of traffic
- Forward proxy and reverse proxy to filter malicious data

Server Hardware Level Protection

The following security measures can be taken at the server hardware level:

- Harden all the production servers to block all unnecessary ports, services, protocols, and software. Remove all unnecessary modules, file shares, filters, and services from the web server and application server.
- Harden the operating system on production servers and remove all unnecessary software and services.
- Install antivirus, vulnerability scanners, and antimalware software on production servers.
- Enable hard disk encryption for the server hardware.
- All production servers should be regularly updated with security patches and security fixes.

Monitoring Infrastructure

Install intrusion detection systems and network monitoring systems. The monitoring infrastructure should continuously monitor the application for security incidents and should report those incidents in real time.

Backup Jobs and Synch Jobs

The system administrators should define backup and synchronization jobs to regularly back up critical data. The files and code should also be regularly backed up.

Disaster Recovery and Business Continuity Plan

In order to fully protect the data, the organization should set up a disaster recovery (DR) environment where data is backed up on a regular basis. During unexpected disasters, the organization can use the DR environment to resume the business within a short span of time.

Implementing Information Security Policies

Defining and implementing robust information security policies are essential to providing robust information security for a DXP. This section discusses various aspects related to information security policies.

Information Access Policies

We need to define security policies and processes to protect information, so that information is appropriately used, distributed, modified, recorded, and destroyed.

- Define policies for sharing sensitive information on social media platforms.
- Conduct security audits regularly to ensure compliance of defined security policies (such as International Organization for Standardization [ISO] and the International Electrotechnical Commission [IEC] 27002). Wherever needed, the organization should also engage an external auditor or certification body to assess security compliance.
- Log, monitor, and track all access change events and admin activities in a secure audit log. The access logs should be retained and archived as per the regulations.

The process of creating an information security policy is detailed in Figure 7-1.

CHAPTER 7 DXP INFORMATION SECURITY

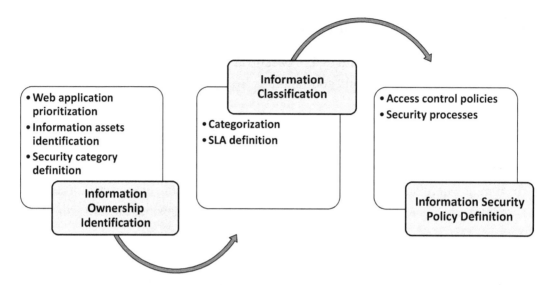

Figure 7-1. Information security policy process

Information Ownership Identification

The first step is to create an inventory of the enterprise applications and prioritize them from a security stand point. For each of the identified applications, identify or create the ownership for information owners. Information owners have the full responsibility of creating or identifying the information assets (such as content, documents, images, videos) and categorizing them based on the sensitive nature of the information.

During this stage we also define various security categories. For instance, we could define three security categories such as public, private, and confidential. The security categories are identified based on their impact. Loss or leakage of public data only causes minor impact, whereas leakage or loss of confidential data leads to huge financial loss and damages reputation. All personal data such as credit scores, date of birth, education status, and such should be categorized into the "private" category.

Information Classification

Once all information assets are identified, they should be categorized into predefined security categories. Each of the information assets are added to one of these security categories. The information owner is responsible for identifying the most appropriate security category for the information asset. The security SLAs for each of the security categories are defined.

CHAPTER 7 DXP INFORMATION SECURITY

The access control list (ACL) should be defined for all resources. By default, the users should be given least privileges to use any resource.

Information Security Policy Definition

For each of the security categories, various attributes such as access levels, distribution and storage policy, and such are defined. Table 7-1 is a sample table that defines a security policy.

Table 7-1. *A sample list of security policies*

Security Concern	Public	Private	Confidential
Access controls	• Given to all • Visitor and guest user access	• Restricted (login- and permission- based) • Maintain technical, physical, and administrative safeguards for the data. • Should not be used apart from the agreed upon purposes. • Autocomplete form fields should be disabled for private data.	• Strictly on need-to know-basis • Multilevel authentication & two-factor authentication that provides additional layer of authentication • Compartmentalized information • Need to sign nondisclosure agreement (NDA) • Confidential information should be masked while displaying. • Confidentiality notice should be displayed on all web pages, reports, screens wherever the data is used.

(continued)

Table 7-1. (*continued*)

Security Concern	Public	Private	Confidential
Storage	• Stored in normal storage	• Access-restricted storage • Ensure continuous availability • Data cannot be cached. • Data cannot be transmitted as URL parameter or with HTTP GET request or as hidden fields.	• Stored in encrypted way • Must be stored in nonreversible one-way hash method. • Stored within a specific location and geography • Ensure continuous availability • Data cannot be cached.
Sharing	• Can be shared as-is	• Need permission from information owner for the storage. • Transport level security • Encrypted during sharing • Bulk sharing not allowed	• Not shared
Destruction		• Should be safely destroyed	• Should be safely destroyed
Auditing and logging	• Not needed	• All access events should be logged.	• All access events should be logged • Third-party audits should be conducted on a regular basis.
Archival and retention	• Must comply with legal regulations	• Need permission from information owner and must comply with legal regulations	• Need permission from information owner and must comply with legal regulations
Availability		• Should have high availability	• Should have high availability

(*continued*)

Table 7-1. (*continued*)

Security Concern	Public	Private	Confidential
Integrity (to prevent information modification)	• Minimal or absence of integrity checks	• Integrity checks should be conducted during data transmission. • During transport and authentication, use certificates created by reputed certificate authorities (CA). • Checksums should be enforced for data.	• Integrity checks should be conducted during data transmission. • During transport and authentication, use certificates created by reputed certificate authorities (CA). • Checksums should be enforced for data.
Confidentiality (to prevent data loss and data theft)	• Not needed	• Should be strictly protected. • Should comply with all security and privacy laws and regulations • Encryption should be enforced by default.	• Should be strictly protected. • Should comply with all security and privacy laws and regulations. • Encryption should be enforced by default.
Incident response		• Inform the impacted users upon data loss or data leakage. • All security incidents should be fixed within 24 hours' time.	• Inform the impacted users upon data loss or data leakage. • All security incidents should be fixed within 2 hours' time.

Protecting Private Data

In order to fully protect the users' private data:

- Identify all the private data of users. This includes PII (personally identifiable information), user preferences, and such.

- Define policies for storing, distributing, access monitoring, and destroying private data.

- Get approval from users when sharing private data with external or third-party party services.

- Define the process for responding to incidents related to breach or theft of private data. The process should identify the roles and responsibilities during a data breach incident.

- Encrypt private data when it is stored or when it is being transferred.

- If the private information is stored in physical records, they should be secured in locked cabinets and should be destroyed at the earliest time.

Information Security Best Practices

This section discusses security-related best practices.

Privacy Best Practices

Privacy information includes PII such as email ID, phone numbers, and such. Privacy information should be transferred only over a secure transport layer (such as HTTPS) and the information should be masked during display. Private information should not be cached and should not be shared with external services. Do not store any private information in session cookies.

Authentication and Authorization

Authentication and authorization should be centrally controlled within an organization. For integration with external third parties, we should use federated security such as SAML. A separate service account should be created for authentication and integration across application layers. A robust password policy should be defined that covers various aspects such as password complexity, password expiration, password storage, account lockout, and such. Simultaneous logins for the same user ID should not be allowed. Use security plugins and filters provided by the platform. Implement the "separation of duties" principle wherein the resource actions are carried out by a separate set of entities. For instance, an application user cannot be an administrator of the same application.

Post authentication, all resources should be provided access based on their roles and permissions.

Auditing and Logging

The information lifecycle events such as creation, updates, and deletes should be logged. All sensitive security transactions such as authentication failures, admin role updates, and password updates should be logged. The audit log entry should include the timestamp, user name, and event details. Private or confidential information should not be included in the log file. Ensure that log files are accessed only by authorized personnel and the log file content cannot be altered.

Engage external security experts to audit the applications.

File Management

The application should only allow whitelist file extensions. Executable files extensions such as .exe, .sh should not be allowed. User-uploaded files should never be allowed to autoexecute and the file permissions should be strictly controlled. Uploaded files should be validated for the file type and size, and scanned for malware or other vulnerabilities.

Error Handling

All security-related errors should be handled. The end user should not see the technical details of the error. For all API invocations and service calls, set a timeout value.

Secure Software Development Life Cycle

A security review should be conducted in every phase of the software development life cycle (SDLC). During the architecture phase and design phases, security requirements should be considered; security reviews and security testing should be carried out during the build phase. Security standards such as SysAdmin, Audit, Network, Security (SANS). Payment card industry (PCI) standards should be followed based on the application and domain needs. Use the latest version of secure open-source components and avoid using open-source frameworks with known vulnerabilities. Check the public disclosures for each of the open-source components used.

Secure Incident Management

Define standard operating procedures (SOPs) to respond to security incidents. The security monitoring infrastructure should be able to immediately recognize security incidents, and the security incident management process should report and address the security breaches. The security monitoring infrastructure should report suspicious transactions. Establish processes and responsibilities to address the security incidents based on its severity.

Database Level Security

Create specific nonadmin database users for the application. Restrict the access to nonapplication schemas and other database packages for the application user. Instead of storing the database user details in plain text, encrypt them or define application server level data sources. Only the DBA user should be allowed to perform operations such as database object creation and modification. Application users should be restricted to do create, read, update and delete (CRUD) operations.

Sharing the Data with External Systems

When the application shares data with external systems and services, it should be strictly based on agreed contracts and should comply with legal regulations. Sharing of private data needs consent from the information owners. The mode of transmission and information loss responsibility should be agreed upon by all parties during the information exchange. The confidentiality, integrity, and availability of the data should be maintained during information exchange.

Security Awareness and Training

All the stakeholders of the organization should be aware of the security processes and policies. In order to achieve that, the organization should conduct mandatory security awareness training for all the employees and they should be made aware of security best practices. In addition to training programs, employees who are handling secure information should undergo mandatory security certification.

Security Testing

As part of application validation, the security testing team should validate all security scenarios for critical business processes and transactions. At a minimum, a security testing process should ensure the following:

- Testing the common security scenarios such as session management, input validation, permission issues, authentication, information disclosure, password policies, header validation, cookie validation, and such

- Validation of OWASP top ten vulnerabilities

- Testing the end-to-end business transactions to exploit any vulnerabilities

- Validation of the security configuration (such as directory browsing) for all servers

- Testing security error handling scenarios

- Conducting automated security testing for testing brute force attacks

- Compiling and reporting out the vulnerabilities to all stakeholders along with recommended remediation actions

- Manual security testing should look for security issues in the logic and do white box testing.

Cloud Testing

Most of the modern digital platforms are available on the cloud or they are cloud enabled. Hence organizations should carefully review the security standards and security controls (such as cloud security alliance and cloud control matrix) provided by the cloud provider. All the controls such as network access controls, resource permission controls, monitoring controls, encryption and key management controls, and such should be reviewed by the security team to ensure that they satisfy the security requirements. The following is a sample list of controls that we can check for while choosing a cloud provider:

- Encryption controls and standards

- Network security support

- Anti-malware support
- Support for role-based access on resources
- Log and remote monitoring
- Security admin controls

Chapter Summary

- Information security defines policies for protecting data at rest and data during transit.

- The key principles of information security are information security policies, data access policy definition, defense in depth, compartmentalization, least privilege by default, and centralized access.

- Defense in depth can be implemented by using infrastructure-level security (firewalls and proxies, server hardware level protection, monitoring infrastructure, backing up jobs and synching jobs, disaster recovery, and a business continuity plan).

- The process of creating information security policy includes information ownership identification, information classification, and information security policy definition.

PART IV

Infrastructure and NFR for the Banking Experience Platform

CHAPTER 8

Quality Attributes and Sizing of the DXP

Nonfunctional requirements, also known as quality attributes, decide the robustness and long-term success of the DXP. The quality attributes such as usability, reliability, scalability, availability, and performance help us to define, track, and measure the success metrics of the digital platform. Conforming to the service level agreements (SLAs) for these quality attributes ensures the long-term success of the program due to higher user satisfaction. This chapter explores various quality attributes needed for a digital solution.

Infrastructure sizing is needed to handle user traffic and transactions for the digital application. We also explore various factors involved in infrastructure sizing and disaster recovery (DR) strategy of a DXP. As the cloud has become the de facto standard for the digital application, we will explore cloud deployment of digital applications.

Key Quality Attributes of DXP

The key quality attributes of DXP are explained in the following list. Each of these quality attributes has its own set of metrics and SLA to quantitatively measure the quality attribute.

- *Scalability*: The ability of the DXP to scale to increased user traffic, data, and transaction volume without compromising the overall performance

- *Availability*: The overall percentage of the time DXP is available and functioning normally over a specific time period (usually 1 year)

CHAPTER 8   QUALITY ATTRIBUTES AND SIZING OF THE DXP

- *Performance*: Performance covers various response-related metrics such as response time, system performance, system throughput, and such.

- *Modularity*: The ability of the DXP to provide and support independent and reusable, plug-and-play type components

- *Extensibility*: The ability of the DXP to support extensions to provide additional future capabilities

- *Security*: The ability of the DXP to protect, manage, and secure sensitive information and data

Other quality attributes relevant for a DXP are as follows:

- *Usability*: Ease with which end user can use the DXP or learn the DXP-based application

- *Accessibility*: The support provided by the DXP to access web applications for people with disabilities

- *Configurability*: The ability of the DXP to allow configuration-driven changes

- *Stability*: The degree to which the DXP can function normally during peak load

- *Interoperability*: Ability of DXP to operate with other systems and services.

- *Efficiency*: The ability of the DXP to perform functions with least resources (CPU, memory etc.)

- *Flexibility*: The ability of the DXP to be reused across various scenarios with minimum modification.

- *Maintainability*: The ease with which the DXP can be enhanced, maintained, fixed, and improved.

In the coming sections, let us look at each of the quality attributes in detail.

Quality Attributes Deep Dive

We will describe various nonfunctional requirements that are typically used for a digital experience platform.

Usability Requirements

Usability defines the "ease of use" for the end user to use the web application and access the required information. The following are the core usability requirements:

- The user interface should support various languages used by users. End users should be able switch the language and to provide the language preference.

- Key functionality should provide contextual help to guide the user on how to use the functionality.

- The system should implement accessibility standards such as Web Content Accessibility Guidelines (WCAG) 2.0 level AA to make the system usable by a wider audience.

- During failures, the system should provide descriptive error messages that are friendly and indicate clear next steps to correct the error.

- Messages should be categorized into categories such as informative messages, error messages, warning messages, and confirmation messages.

- The website should provide consistent user navigation (menus, breadcrumb) and provide uniform and consistent page layouts.

- The system should provide robust online help providing contextual help and in-context hints.

- The website should provide how-to guides, usage guides, FAQs, product documentation, and tutorials and manuals.

Security Requirements

Security details are detailed in Chapters 6 and 7. The following is a look at prominent security requirements:

- Ensure that all private web resources are properly protected.

- The system should provide strong password policies: that includes strong passwords, password change frequency, account lockout policies, and such.

- Account-related operations such as profile updates, password changes, and registration should provide strong authentication mechanisms (such as reauthentication, multifactor authentication, step-up authentication, adaptive authentication, etc.).

- Default admin passwords should not be used at any layer.

- Forgot password functionality should send a time-sensitive password reset link instead of sending the password in plain text.

- The session should be invalidated after explicit logout or after a specified inactivity period. Session IDs should be sufficiently random, and the application should not permit multiple simultaneous sessions.

- Authenticated sessions should use HttpOnly cookies (cookies that are cannot be accessed by client side scripts) and they should use a "secure" attribute and a strict transport security response header that instructs the browsers to access the resources only over HTTPS.

- All direct object references that provide the direct handle to internal objects or data blocks should be protected. Authenticated users should only access authorized data and objects.

- The filenames or folder names obtained from untrusted sources should be canonicalized (converted into proper file name or folder path) before using them.

Reliability Requirements

Reliability ensures that the system consistently performs the specified functionality and recovers from failure. Given below are key reliability requirements:

- The system should consistently provide expected behavior.

- If any of the back-end systems or service fails, the digital application will not fail. The application should be able to gracefully recover from the failure.

- The system will handle the re-initiation of operations that lead to an inconsistent state. For instance, the system will prevent double clicking of the "checkout" function before the first checkout operation is completed.

- The system handles unexpected and error scenarios such as long-running transactions, failed transactions, network errors, outage of back-end systems, and such. The system logs the issues and gracefully rollbacks and shows friendly error messages in such scenarios.

- The infrastructure should be designed to be fault tolerant.

Scalability Requirements

Scalability ensures that the application scales well for the larger workload, with acceptable performance. The following are the main scalability requirements:

- For on-premise deployments, the application should be deployed as a multinode cluster.

- The system should be able to handle increased user load and data load, and should be able to handle future data growth.

- Horizontal and vertical scaling should be supported by the system.

- If any of the cluster nodes goes down (web server node, application server node, or database server node), the system should be able to load balance and seamlessly switch all the traffic to other nodes to provide the expected response.

- The system should provide response within an acceptable time period, even during peak traffic.

Availability Requirements

Availability provides the uptime availability of the system. Maximum availability is necessary to ensure high-quality service for the end users.

- The system should provide maximum and continuous availability. In order to provide preferred 99.999% availability, the system downtime should be a maximum of 5.26 minutes per year.

- Even during peak traffic, the system's availability should not be impacted.

- The availability of the system should be ensured even when one of the back-end systems or services is down.

- The data, code, and content should be synched up on a regular basis to a disaster recovery (DR) site to handle unexpected disasters and ensure business continuity.

- In case of unexpected disasters, the DR site should be up and running within the recovery time objective.

- The recovery time objective (RTO) (the maximum time period within which the DR site should be made active during disasters) should be designed based on availability SLAs. In order to provide 99.999% availability, we should have an active DR site that can take over immediately after the primary site is down.

- The recovery point object (RPO) (maximum period for which data loss is tolerated) should be designed based on availability SLAs. In order to provide 99.999% availability, we should have an active DR site to which we do a real-time data sync.

- The system should be available during upgrades and patching process.

Archival and Retention Requirements

Due to legal obligations or business needs, content and data need to be archived and retained for a specified time period. The main archival and retention requirements are as follows:

- The system should archive and retain all data and content to comply with legal regulations.

- The system should store the data in separate storage.

- The system should be capable of archiving data for a specified time duration.

- The archival system should provide data redundancy to prevent data loss and to provide high availability.

Logging and Auditing Requirements

Logging is necessary to understand and debug the system actions. Auditing is necessary to log security events such as login, login failures, login attempts, password change, etc. The main logging and auditing requirements are as follows:

- The system should log the application events (with appropriate categories such as info, debug, or error) and security events.

- The log entry should consist of timestamp, event source, error details (if any), source IP address, and user ID so that it helps the administrators to use the information.

- Security events such as login failures, login attempts, elevation of privileges, account registration, workflow approvals, etc. should be logged with corresponding user IDs.

- Sensitive information such as private information and session ID should not be logged.

- Appropriate authorization and access controls should be defined for the log files.

- All sensitive operations such as creating a user, deleting an account, or updating an account should be persisted in the audit table. The audit table should store the transaction ID, user ID, timestamp, old value, new value, and the operation type (create, update, delete).

Performance Requirements

A system's performance is measured by metrics such as response times, page load time, etc. Following are the main performance requirements. We discuss DXP performance in detail in Chapter 9.

- The response time of web pages should be within 2 seconds across all geographies and all access channels during average load. In order to fulfill strict SLAs across all geographies, we may need to use content delivery network (CDN) and geo-centric applications.

- If the system takes more than 5 seconds for any transaction, the system should display an informative message or graphical icon to the end user.

- The system should respond within accepted SLAs even during peak traffic and during maximum concurrent transactions.

- The throughput for the web server and application server should be able to support maximum transactions and page views.

Infrastructure Sizing of DXP

A properly sized infrastructure is the most essential element for achieving optimum scalability and availability for the platform. This section looks at the key factors that can be considered for infrastructure sizing of a DXP.

Note The following sizing calculation is for on-premise deployment.

Table 8-1 provides the key factors that are used in DXP sizing:

Table 8-1. *Sizing Metrics for DXP*

Category	Metrics needed for sizing
Load numbers	• Total number of application users • Maximum number of anonymous/guest users. • Maximum concurrent users • Maximum transactions per hour • Load number growth rate per year
Session numbers	• Average session time per each user • Maximum objects per user session • Average size of each session object
Content numbers	• Maximum web content volume • Need for separate authoring and publishing instances • Content number growth rate per year
Availability number	• Application uptime requirement • Disaster recovery requirement • Availability SLAs
Performance numbers	• Average page response time • Performance SLAs
Throughput numbers	• Average number of page views per person • Average page size • Maximum number of visitors per day

Given as follows are sample high-level calculations.

The sample calculations to calculate the right sizes for RAM and CPU cores for a server is shown as follows.

- Maximum RAM memory needed by the application (in KB) = (maximum number of concurrent user sessions × Average session size) / 1024.

- Maximum number of CPU cores needed = (maximum number of web page requests per second) / (average number of pages served by a single core in 1 second) + (maximum number of resources requested by users per second) / (average number of resources served by a single core in 1 second).

- Minimum bandwidth needed per day = (average number of page views per person per day × average page size in MB × maximum number of visitors per day).

Cloud Hosting of DXP Solution

Modern digital platforms are available as cloud native applications or they provide a cloud deployment option. Cloud native applications and cloud-enabled applications help organizations to optimize cost, enhance business agility, reduce deployment times, and provide highly available and secure services. The cloud enables rapid prototyping and faster innovation. In this section we will discuss the main factors needed for cloud deployment of a DXP solution.

Tiered Architecture

We need to identify all the components for implementing the tiered architecture. At a minimum we should configure the following components:

- *Firewall*: This component is needed to filter the traffic and prevent any network-related security threats.

- *Load balancer*: Load balancers are used to evenly distribute requests to all available systems.

- *Security providers*: We need to configure the security components to provide authentication services.

- *Web server*: The web server caches and serves the static content such as images, videos and other binary content.

- *Application server*: The enterprise application is hosted on the application server.

- *File storage server*: The server is used to store the file and content.
- *Database server*: The application data will be persisted in the database server.

Depending on application needs, we also need to configure the e-mail server, content management server, authentication server, and document management system.

We need to use the load requirements and traffic requirements to arrive at the appropriate infrastructure sizing numbers for each of the tiers. For high availability, we need to use multinode cluster topology.

Following are other factors that we need to consider for cloud deployment:

- *Availability*: Most cloud providers offer high-availability deployment models. This includes multiregion availability, fault tolerance, and disaster recovery options.
- *Scalability*: Auto scaling is one of the core strengths of the cloud model. We need to check the cloud provider's scalability needed for the application and plan the cloud deployment.

Cloud Deployment Considerations

If we are planning to deploy on-premise applications to the cloud, we need to evaluate the suitability and feasibility of the cloud deployment. Following is a list of evaluation parameters that we can consider for cloud deployment.

Platform Coexistence

Normally, enterprise applications have integrations with internal/legacy enterprise applications. When we move the enterprise application to the cloud, we need to evaluate the options for these integrations. Following are some of the factors we need to evaluate:

- Move the enterprise application along with all dependencies to the cloud. If there are strict data and security requirements, evaluate the option of connecting from cloud to on-premise secure systems or using a private cloud.

CHAPTER 8   QUALITY ATTRIBUTES AND SIZING OF THE DXP

- Use a virtual private cloud (VPC) to integrate with in-house/on premise applications.

- Leverage adaptors provided by the cloud provider to integrate the enterprise application to the on-premise applications.

Security

Security involves various categories, as discussed in Chapters 6 and 7. From a cloud deployment standpoint, infrastructure security and data privacy are key concerns. We need to carefully evaluate the infrastructure-level security provided by the cloud provider; that includes firewall, layered security, and security measures against denial of service (DoS) or distributed denial of service (DDoS).

We should also check for regulatory requirements and compliance requirements for the data storage location and options provided by the cloud provider for the same.

Integration design

Most of the modern digital applications use service-oriented design for integrations. The digital platform should be designed to use REST-based services for all its integrations. This would provide an easier and extensible way to integrate with all external and third-party applications. When we deploy the CMS or other systems in headless mode, the RESTful integration model can easily consume (and is flexible to adapt to) this headless integration model.

Cloud Deployment Model

A typical cloud deployment model for a DXP application on the Amazon Cloud is shown in Figure 8-1.

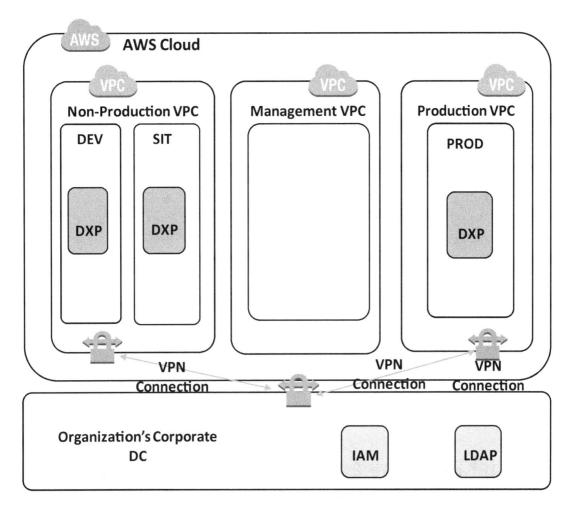

Figure 8-1. *Sample DXP Cloud Deployment Model*

Key highlights of the cloud deployment model of the DXP application as depicted in Figure 8-1 are as follows:

- The environments will be hosted in the specific region of Amazon Web Services (AWS).

- There are three environments Prod, SIT, and Dev; these will be hosted in two separate VPCs, designated as Prod and Non-Prod VPCs.

- A separate management VPC will serve both the Prod and Non-Prod VPCs. In Figure 8-1 we have "Production VPC" for running the production instance of DXP and "Non-Production VPC" for running the Dev and SIT instances of DXP.

- Each VPC has been segregated into tiers by creating separate subnets for each tier.

- Connectivity to the back-end services like LDAP and IAM that are hosted on-premise will be secured via an IP Sec VPN. There will be two VPN's set up: one for Prod and the other for the Non-Prod VPC.

- Customers and end users will access the DXP application over Secure Sockets Layer (SSL).

- A multiavailability zone configuration is recommended for production, as this provides zone level failure and automated failover within seconds.

- All nonproduction workloads will be deployed as standalone, and production workloads will be deployed with high availability.

Note We have taken Amazon Cloud as an example to depict the cloud deployment model. However, we could use any cloud provider based on the requirements.

Disaster Recovery and Business Continuity for DXP Applications

Disaster recovery (DR) ensures high availability during unexpected events such as a natural disaster. A robust DR strategy is needed in ensuring business continuity. This section elaborates the DR process for DXP applications.

The high-level steps in a DR strategy are given in Figure 8-2.

CHAPTER 8   QUALITY ATTRIBUTES AND SIZING OF THE DXP

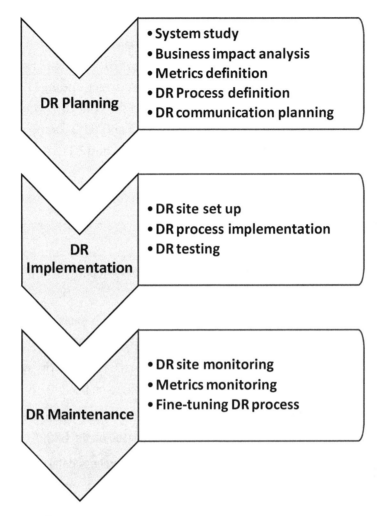

Figure 8-2. Sample DR strategy

A DR strategy includes three main steps, as depicted in Figure 8-2: DR planning, DR implementation, and DR maintenance.

DR Planning

Disaster recovery planning involves identifying all DR data and DR processes to implement a robust DR strategy. We start by studying the as-is system through stakeholder interviews and questionnaires to understand the code, content, and data that needs to be synchronized on a frequent basis. During system study, we identify the system dependencies, understand system and network architecture, and study the historical outage incidents. We should carry out a business impact analysis for critical business processes; the impact analysis exercise will help us to map each of the business-critical processes to the component and DR process. The business impact analysis also helps us to define the RTO for the corresponding DR process. We should also examine existing contingency policies and data backup processes. During this phase we will also finalize the metrics such as RPO and RTO for the site. We will define the DR activities and processes to achieve the RPO and RTO. A sample list of DR activities is given in Table 8-2.

Table 8-2. Sample DR Activities

DR Activity	Process	Comments
DR site setup	Site setup process	Create a DR site as a mirror replica of the primary site as a one-time setup process.
Code backup	Code backup process	Code is synchronized on a daily basis. The synchronization frequency is adjusted based on the RPO.
Data backup	Data backup process	Data is synchronized on a daily basis. The synchronization frequency is adjusted based on the RPO.
Content backup	Web content backup process	Content is synchronized on a daily basis. The synchronization frequency is adjusted based on the RPO.
DR Site switching	Site switching processes	The emergency response process details the steps to switch from primary site to DR site.

The infrastructure architect defines the governance model and details all the required infrastructure-related processes. Most of the synchronization processes will be configured as scheduled batch jobs. The infrastructure architect reviews the metrics, activities, and processes with all stakeholders and gets their signoff. The infrastructure architect also defines the communication process for internal and external stakeholders as part of DR process definition. Once the DR processes and activities are finalized, the infrastructure architect defines the schedule for implementing the planned DR activities.

DR Implementation

During the implementation phase, we assign the resources for each of the planned activities based on the schedule and priority.

The primary activity is to set up the DR site in a remotely located data center and set up all the synchronization jobs. The infrastructure configuration at the DR site should be an exact mirror replica of the primary site. The DR process owners set up and configure all the backup and synchronization jobs as part of setup activity.

We will prepare a detailed DR test strategy to test all the DR processes and synchronization jobs. We will execute the pilot run of DR processes and measure the RTO and RPO to ensure that DR processes and their frequency are well designed.

There are mainly three types of DR options available, as listed in Table 8-3.

Table 8-3. *Key DR options*

DR Option	Setup and Synchronization Details	Typical RPO and RTO
Cold backup option	The DR site is prepared and set up but the data and content are not actively synched. The data and content will be restored from backup in the event of disaster.	RPO = 24 hours RTO = 72 hours
Warm backup option	The DR site is synced with data from the primary site in regular intervals. In the event of disaster, the warm backup takes over.	RPO = 4 hours (with 4-hour sync jobs) RTO = 1 hour
Hot backup option	The DR site is synced with data from the primary site in near real-time. In the event of disaster, the hot backup takes over.	RPO = 1 hour (with hourly sync jobs) RTO = 30 mins

The DR options can be chosen based on RPO and RTO needs. The hot backup option is used for very high availability requirements and mission-critical applications that have strict RPO and RTO. The cold backup option can be used for applications that can tolerate data loss or low availability.

DR Maintenance

The DR processes are continuously monitored and their success/failure reports are notified to all stakeholders. Based on the changes in business objectives, the DR processes are fine tuned. The DR processes and plans are reviewed and updated at regular intervals. All DR teams are trained on the emergency operating procedures.

DR Strategy Document

What follows is the structure of a DR strategy document covering all concerns and topics related to the DR process.

Scope and Objectives

This section mainly defines the DR requirements and covers these elements:

- Scope of DR planning
- DR objectives
- Assumptions

As-Is System Analysis

In this section we will understand the existing data, processes, and system architecture to come up with the DR process and architecture.

- System architecture study
- Metrics definition: RPO and RTO
- Business process study and prioritization
- Key disaster scenarios

DR Planning

The main activities in DR planning are given as follows:

- DR roles and responsibilities
- DR processes
  - Synchronization process
  - Communication process
  - Incident response process
  - Site switching process
  - Recovery process for each disaster scenario

- Monitoring activities
  - Primary site monitoring
  - DR site monitoring
  - Metrics monitoring and notification

Chapter Summary

- The key quality attributes of a digital application are scalability, availability, performance, modularity, extensibility, and security.
- Other quality attributes are usability, configurability, stability, interoperability, efficiency, flexibility, and maintainability.
- Usability involves supports for various languages, contextual help, accessibility, friendly error messages, FAQs.
- Security covers various aspects related to authentication, authorization, password policies, and such.
- Reliability requires consistent performance of the application.
- Scalability requirements are met if the application scales for increased user load and transaction load.
- Availability requires continuous uptime and availability of the application.
- Performance requirements cover application response time, process time, transaction completion time, and such.
- Infrastructure sizing includes various factors such as load numbers, session numbers, content numbers, availability numbers, performance numbers, and throughput numbers.
- A robust DR strategy includes DR planning, DR implementation, and DR maintenance.

CHAPTER 8 QUALITY ATTRIBUTES AND SIZING OF THE DXP

- DR planning includes system study, business impact analysis, metrics definition, DR process definition, and DR communication planning.

- DR implementation includes DR site setup, DR process implementation, and DR testing.

- DR maintenance includes DR site monitoring, metrics monitoring, and fine tuning the DR process.

CHAPTER 9

DXP Performance Optimization

Performance is the primary driver for end-user satisfaction. User-centric DXP platforms aim to provide optimal performance at all touch points. Performance also directly impacts user traffic, user retention, online revenue, and conversion rate. Page performance is also used by some web search engines for ranking the web page in further improving the site traffic.

This chapter discusses various dimensions of DXP performance optimization and various methods that are used to achieve the same.

DXP Performance Optimization of Presentation Layer

In many scenarios, performance is taken up during the end phases of the project when the team notices performance delays during performance testing. Troubleshooting and fixing performance issues during the end stages of the project is not only costly but also impacts the project timelines. Performance optimization should be done iteratively to identify and address performance issues during early phases of the project.

User Experience

End-user experience is of paramount importance for a DXP. As web pages become more interactive, the page size and associated performance overhead are also increased. Enhancing the user experience involves optimizing the performance at all user touch points.

CHAPTER 9 DXP PERFORMANCE OPTIMIZATION

Web Page Performance Optimization Scenarios

Most of the user journeys are initiated through web pages, hence it is necessary for the web pages to have optimal performance. Table 9-1 lists rules of thumb and performance best practices for web page optimization.

Table 9-1. Performance Optimization Thumb Rules for Web Page

Performance Optimization Rule of Thumb	Objective	Techniques
Reduce HTTP requests	• The browser spends time in establishing connection and downloading the components. Reduced HTTP requests minimize connection and download time.	• Merge JS/CSS files into a combined master file. • Use CSS Sprite. • Remove any duplicate HTTP requests.
Image optimizations	• Rendering images takes most of the page load time. • Delivery of optimal version of image greatly reduces the page load time and reduces overall page size.	• Compress the image to reduce its size. • Use optimal image formats such as PNG or WebP • Progressively load images on demand. • Specify exact image dimensions.
Optimal resource loading		• Load the resources asynchronously. • Use responsive images that renders on all devices and browsers. • Defer JavaScript loading.
Caching	• Caching frequently used data improves the performance	• Set cache control headers for static assets such as images, Javascripts, and CSS files. • Cache AJAX responses for faster rendering.

(continued)

Table 9-1. (*continued*)

Performance Optimization Rule of Thumb	Objective	Techniques
Delivery from edge locations (nearest geographical location)	• For static content, leverage globally distributed edge locations (nearest geographical location) to serve the content with minimal latency. • Deliver content from the most optimal geographic location.	• Use content delivery networks (CDNs) like Akamai. • Use cloud-based delivery for static content.
Compress and reduce size of the presentation components.	• Compression reduces the overall request size and the overall page size. • The size of text-based components such as JavaScript files and CSS files reduces after compression.	• Enable gzip compression at the web server layer and set appropriate HTTP headers for browser to decompress. • Minify the JavaScripts by removing unnecessary and redundant data and stylesheets and plugins.
Optimal positioning of the JavaScripts and stylesheets	• Appropriate placement of stylesheets and JavaScripts improves the perceived page load times and prevents browsers getting blocked.	• Place CSS at top and JavaScripts at bottom. • Deliver JavaScripts from CDN and cloud locations.
Other performance optimization rules of thumb		• Avoid server redirects. • Use custom build of JS frameworks to include only the needed modules/components. • Remove duplicate HTTP calls. • Fix all 404 errors and avoid 301 redirects. • Use HTTP/2, which is the latest version of HTTP. • Use multiple domains to host and deliver static content.

(*continued*)

Table 9-1. (*continued*)

Performance Optimization Rule of Thumb	Objective	Techniques
Mobile web performance		• The mobile web application could save the JS/CSS in the device's local storage and use it for subsequent requests. This optimizes latency. Use the offline storage features of the mobile platforms. • For second time page loads, only load the updated/modified scripts not stored in local storage.

Performance Testing for DXP

We need to periodically test the performance of web pages. Performance testing tools identify the performance-related issues and provide the recommended action plan.

Performance Testing Activities

Key performance testing values for a regular sprint-based delivery are depicted in Figure 9-1.

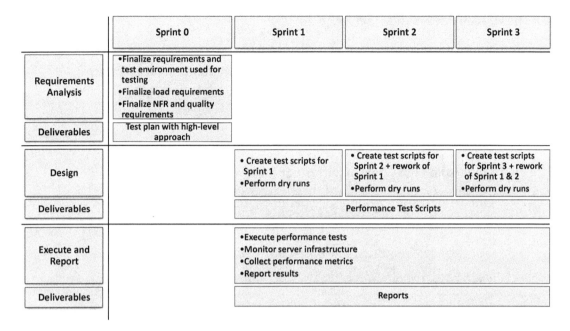

Figure 9-1. *Various sprints for performance testing*

We have elaborated categories of performance testing as follows:

Requirement Analysis

During requirement analysis, we identify the detailed requirements related to performance testing. The requirements are obtained through stakeholder interviews and requirements documents. Requirements analysis is done in sprint 0 so that the test plan will be used in subsequent sprints. Chapter 2 covers requirement analysis in more detail.

The main objectives during this stage are:

- Identify performance test requirement/expectations.
- Identify all the performance service level agreements (SLAs) for page response times, transaction completion time, and such.
- Identify and finalize load numbers such as maximum concurrent users, average session time, maximum content size, and such.
- Define an approach for performance test execution.

The key activities in this stage are:

- Interview stakeholders and gather performance test requirement/expectations. Other requirement gathering methods are as-is system analysis, and study of requirements documents that can be used.
- Understand risks, constraints, and assumptions related to performance.
- Understand the scope of performance testing for each sprint.
- Understand application usage patterns, traffic patterns, content volume, and such.
- Prioritize test requirements based on business criticality.
- Understand data setup requirements and data quality requirements.
- Prepare a high-level performance test plan for the sprint.
- Prepare a test plan based on the requirement analysis and stakeholder interviews.
- Conduct a review of the test plan to verify the performance SLAs and performance expectations.

Design

During this phase we design and develop the performance test scripts as per the test plan prepared in sprint 0. The performance test scripts can be categorized as client-side performance test scripts and server-side performance test scripts.

The main object of performance testing in this stage is to create the test scripts to provide coverage, and execute the performance test scripts.

The key activities in this stage are as follows:

- Set up the test environment and set up test data.
- Create performance test scripts as per the test plan.
- Set up a continuous and iterative testing environment.
- Set up a test reporting and notification infrastructure.
- Set up parameters to be monitored at the infrastructure resource level.

Performance Testing Execution and Reporting

During this phase we execute the performance test scripts and report the findings. The main objective of this phase is to verify all test scenarios to provide a stable build.

The key activities in this phase are as follows:

- Refine test scripts and test data as required.

- Perform identified types of performance tests like load, stress, endurance, etc.

- Ensure required workload is being generated at various points.

- Execute performance tests to cover various scenarios:

  - Isolated tests at different workloads for individual transaction. This helps us to test the performance behavior of each of the transactions.

  - Mixed load test at different workloads. We can test this by adding various loads for a combination of transactions.

- Ramp-up tests that incrementally add user load in incremental steps. Work load and test duration is gradually increased, and the server resources and response times are monitored. We can increment the user load in steps of 10 users per test.

  - For each transaction such as checkout function or shopping cart function

  - For mixed transactions such as a combination of checkout and shopping cart

A sample table to record the ramp-up testing is shown in Table 9-2.

CHAPTER 9   DXP PERFORMANCE OPTIMIZATION

Table 9-2. Sample Ramp-Up Test Table

Work Load	Time Duration	CPU Utilization	Memory Utilization	Throughput	Page Response Time
10	10				
20	20				
30	30				
40	40				
50	50				

We can perform the ramp-up test as mentioned for each of the key transactions, to understand the system behavior at each load. We could also perform the ramp-up test on a transaction mix (including a combination of transactions).

- Monitor and capture performance statistics of server resources (CPU, memory, network bandwidth) and infrastructure.
  - Generic parameters:
    - *CPU utilization*: total utilization, idle time
    - *Memory utilization*: committed bytes, available bytes, etc.
    - *Page performance*: page response time, perceived response time, time to first byte
    - *Physical disk*: read, write, and latency
    - Throughput
    - Instrument code with timers for method / code level timing
  - *Server side parameters*: connection pool parameters, thread pool parameters, etc.
  - Network monitors
  - Profiling, code coverage, and memory debuggers

- Collect and report test results:
  - Response times vs. number of users
  - Throughput vs. number of users
  - Transactions per second
  - Hits per second
  - Users vs. time
  - Errors
  - CPU utilization vs. number of users/time
  - Memory utilization vs. number of users/time
  - Throughput achieved

We can document the response times at various workloads as shown in Table 9-3.

Table 9-3. Ramp-Up Test for Transactions

Workload	Transactions	Response time			
		Minimum Time	Average Time	Maximum Time	90% Time
100 users	User login				

Key Performance Metrics

The main performance metrics for web page components are given as follows:

- *Time to first byte (TTFB)*: It is the time taken by the server to send the first response to the browser. TTFB is the measure of server responsiveness. A smaller TTFB is essential for optimal page response time.

- *Page size*: It is the overall size of a web page including the HTML and static assets (images, videos, scripts, stylesheets, fonts, etc.). A large page size delays the page load time, hence it is recommended to reduce the overall page size.

- *Page response time (PRT) or page load time*: This is the total time taken for the web page to load on the user agent/browser. The general user expectation is to load the page within 1 second (for HTTP pages) to 5 seconds (for HTTPS pages).

- *Above the fold time*: It is the time taken to render the page components within user view (above the fold).

- *Perceived response time*: It is the page load time perceived by the end user. We should aim to keep a minimal perceived page load time for improved user experience.

Performance Testing Framework

Various elements of the performance testing framework are depicted in Figure 9-2.

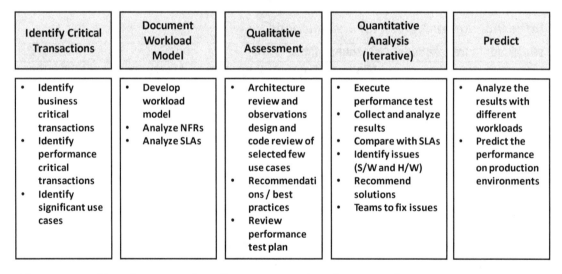

Figure 9-2. Key elements of performance testing framework

Various elements of the performance testing framework are elaborated as follows.

Identify Critical Transactions

In this phase we identify the business-critical transactions. The following are the key activities in this stage:

- Understand architecture and design.
- Review of existing systems and infrastructure elements.
- Understand nonfunctional requirements (NFRs) and SLAs.
- Understand performance pain points in the current system.
- Identify business-critical transactions for the application.
- Identify performance-critical transactions and collect data required for the workload model.
- Select key use cases for design and performance-centric code review.

Document Workload Model

In this stage we model the workload for the application. Based on continuous monitoring and log file analysis we identify the user load (such as maximum users, concurrent user, peak user traffic, average user think time, logged in vs. anonymous users), content load (maximum content volume, content growth rate), page access patterns (frequently visited pages, popular navigation path), session values (average session time, session size), and such. We document all the key NFRs (such as availability, performance, scalability) and the SLAs for the same.

The workload values will be used for performance test scripts and for load testing.

Qualitative Assessment

During the qualitative assessment phase, the performance engineer analyzes the as-is system (such as infrastructure, server configurations, code, etc.) and provide best practices-based recommendations to address any identified performance issue. During this phase, the performance engineer also reviews the performance test plan.

CHAPTER 9 DXP PERFORMANCE OPTIMIZATION

Quantitative Assessment

During this phase the performance test scripts are executed, and the performance engineering team monitors the system behavior and the following server infrastructure components.

1. Resource level

    a. Utilization (CPU, disk, memory, network)

    b. Throughput

    c. Response time

    d. User load

2. Web server

    a. Maximum threads

    b. Keep alive connections

3. Application server

    a. Execution threads

    b. Object pooling

    c. Execute queue length

    d. Entity beans pool

    e. JDBC connection pooling

    f. Garbage collection

    g. Response time of servlets

4. Database server

    a. Buffer cache hit ratio

    b. Redo logs, top 5 SQL statements

    c. Indexes

    d. Buffer waits

Performance testing results will be compared against the specified SLAs. All issues identified in performance testing will be addressed.

CHAPTER 9 DXP PERFORMANCE OPTIMIZATION

Predict

Performance testing will be conducted for various workloads; using the test results, the performance in production environment will be predicted.

Performance Debugging Framework

When we encounter performance issues in DXP applications, we troubleshoot the root cause of the performance issue. Generally, troubleshooting performance issues is a complex exercise because it involves all the layers and components in the web request processing pipeline.

In this section we discuss a performance debugging framework that provides proven methodical steps for troubleshooting performance issues.

The main steps of the performance debugging process are as follows:

- Step 1: Identify the root cause component/system.
- Step 2: Fix and Optimize the component causing the performance issue.
- Step 3: Perform load testing, peak testing, and stress testing to ensure that a fix done in Step 2 works optimally in all scenarios.

Let's discuss each of these steps in detail as follows.

Identify the Root Cause

We need to check the performance at each layer to understand the component causing the performance issue. This is often the most complex step in the process. We need to devise test methods for each layer and for components involved in the web request processing pipeline. In addition, we need to do runtime profiling of the application and log analysis to get more insights into the problem-causing element.

Following are various methods to identify the performance root cause:

- Perform layer-wise performance testing. Identify the time taken at each layer involved in the web request processing pipeline. Starting from the presentation layer (Java server pages/active server pages) to the database layer, get the average execution time (for more than ten iterations). At a minimum, we need the execution time for following:
    - Average load time taken by the overall web page

- Average time taken for each of the page components (JS, CSS, images, header footer, widgets, video, etc.): Record the page load time, perceived load time, asset load time, page size, and other key performance metrics.

- Average time for server call on each web page: Identify the Time to First Byte (TTFB) to understand the server response time.

- Record the performance metrics at various loads to get average numbers.

- Average time for server side components:
  - Once we get insights about the TTFB for all the web pages, we understand the time-consuming server component.
  - Once the server component is identified, we need to profile the component to further troubleshoot the issue.
  - We need to calculate the time needed for various calculations, service calls, and database calls that are involved in the execution of the component.
  - Increase the load and retest the preceding scenario to get the average execution times.

- Average time spent at database layer/services layer:
  - Based on the profiling of the server-side component, if the performance issue is traced to the database layer or services layer, identify the problem-causing query or service call.
  - Execute the time-consuming query and identify the performance optimizations needed for the query (using an explain plan and other database-related optimizations such as indexes, query caching, etc.)
  - Identify the performance optimizations for the service call (such as using server-side caching, optimal sizing, fine tuning server configurations, etc.).
  - Increase the load and retest the scenario to get average times.

- Record the time for various steps in the business process and transaction:
  - For critical business process and time-consuming transactions, record the time taken for each step in the process.
  - Repeat this step by increasing the load.
- Profiling the components:
  - Profiling presentation layer:
    - In addition to layer-wise performance metrics, use developer tools to profile the web pages.
    - Web page profiling provides useful metrics such as categorized asset size, categorized asset load time, and such.
  - Profile the server-side components to understand the memory consumed by server side components:
    - Business components performance analysis
    - API and call tracing analysis
    - Database call performance analysis
    - Enterprise integration performance analysis
  - Profile the database by understanding the execution plan for the queries:
    - Review the data model to see if it is properly utilized.
    - Explore the possibilities of using lookup tables and snapshot tables to prepare/cache the data for frequently accessed queries and for static values.

- Log analysis:
  - Analyze the logs to see if there are any exceptions/errors/waits/ deadlock or any other obvious performance issues. Try to get the execution timings (from logs/data from tables) to find out the timings from the application tier (request received time, response sent time, response size, etc.)
  - The DBA can analyze the logs on the database end.
- Server configuration analysis:
  - Server administration needs to check all configurations of server to check if all the settings/configuration are optimized as per recommended best practices. The settings include, but are not limited to: connection pool settings, Java virtual machine settings, thread settings, log configurations, cache settings, heap settings, session settings, and any other application server-specific settings recommended by the product vendor.
  - The DBA can analyze the database server settings to ensure if all parameters are properly configured.
- Infrastructure/capacity analysis:
  - Infrastructure experts need to verify the existing infrastructure sizing (CPU, memory, disk size, storage, network bandwidth, cache server, CDN, etc.) to check if it is optimal to support the required performance SLAs, content load, and traffic needs.
  - Network experts need to analyze the network traffic to understand its extent.
- Ramp-up testing:
  - For all performance testing, steadily ramp up the user load starting with 50 users for 30 minutes and then increase the user load to 100 users, 150 users, and so on as per the traffic requirements.
  - This steady ramp test executed over longer duration reveals the performance issues, memory leaks, and breakpoints.

Optimize the Component/System/Layer

For the first step we will identify the root cause for the performance issue. Once the problem-causing component is identified, the next step is to optimize it. Following are the generic guidelines for performance optimizations.

- Fine tuning the database:
  - Use materialized views to store the database query results for popular queries.
  - Use database hints.
  - Optimize queries and avoid Cartesian join (a join of each row of one table to each row of another table).
  - Check if the indexes are created and used by the application queries.
  - Calculate the DBA statistics for the key tables.
  - Check the explain plan for the cost of the query and fine tune the query based on the explain plan.
  - Wherever possible, aggregate functions or stored procedures can be used to do the database heavy lifting operations.
  - Check if there are any "full table scans" (a query resulting in a scan of every table row) happening instead of an index scan (a query result using the indexed columns).
- Fine tuning a server:
  - Turn off logging.
  - Carry out calculations in the database.
  - Use connection pooling (reuse of existing connections through a managed pool) wherever possible.

CHAPTER 9 DXP PERFORMANCE OPTIMIZATION

Common Performance Problem Pattern

In this section we discuss the common performance issues observed in DXP applications and their root causes.

Following are some of the commonly encountered performance issues that we have seen in the past:

- *Performance issue*: Applications not scalable. Some symptoms, with increase in load:

    - Response time increases drastically.

    - Some queuing is observed.

    - Gradual performance degradation with increase in user load

    - CPU utilization at some layer remains constant.

    - *Common reasons*:

        - Absence of caching framework that caches frequently used objects

        - Existence of single point of failures

        - Absence of clustered setup for web server, application server, and database server. A clustered setup involves multiple nodes/machines to serve the response providing failover and high availability.

        - Inappropriate business object tuning. If the business objects are not tuned for performance, it has ripple effect on the overall application's performance.

        - Inappropriate infrastructure sizing such as CPU cores, memory, network bandwidth, and such.

        - Inappropriate connection pool settings such as maximum connections, minimum connections, connection idle time out, and such

- *Recommendations*:
  - Minimize session size by storing only absolutely needed objects in the session.
  - Size the infrastructure to handle the maximum user load.
  - Use multinode clustered topology for all servers to eliminate single point of failures. A multinode topology involves using multiple nodes in a setup to handle user requests.
  - Use a caching framework or caching server (a dedicated server to cache frequently used objects) to cache frequently used data and lookup data.
  - Configure the application server parameters to handle maximum user load. The common parameters for Java-based application servers are given below:
    - Garbage collection metrics
    - Tuning parameters
      - Xms, Xmx
      - NewSize, MaxNewSize
      - PermGen, MaxPermGen
- *Performance issue*: Inappropriate deployment architecture leading to single point of failure. Few common symptoms given below:
  - Multiple applications were deployed on the same server in production and pre-production environments. Performance testing revealed that this setup was not scalable.
  - CPU utilization graphs showed spikes during normal load testing. This was attributed to problems with one of the applications, which was not scalable.
  - Applications, admin functionality, and search modules shared the same server. This led to performance issues in search and admin functionality.

- *Common reasons*:
  - Because of the nonscalable behavior (performance degradation with increase in user load) of the application, other applications are not able to support the projected volumes with this setup.
- *Recommendations*:
  - Move the nonscalable application to a separate server. On a separate server this application was able to support the projected volumes. Alternatively transform the application to use microservice architecture and independently scale the microservices.
  - Do profiling of the nonscalable application to identify the root cause of the issue.
  - Deploy application, admin modules, and search modules on separate, individually scalable servers and containers.

Performance Case study

This section reviews a performance optimization case study for a DXP-based application.

Application Context and Background

The DXP application was built for a manufacturing organization. The DXP application provided web interfaces for suppliers, administrators, and distributors. During the user acceptance testing (UAT) testing, the team noticed performance issues on the landing pages for suppliers and distributors. The landing pages provided functionality to view the payments, processing, and transaction details for suppliers and distributors. The landing pages heavily used database queries to get the matching results. The landing pages were taking more than 120 seconds to load on various supported browsers.

Performance Analysis

Performance engineers applied the performance debugging framework we discussed earlier. Performance engineers analyzed the slow performing pages end to end to analyze all the layers and components involved in the web request processing pipeline.

CHAPTER 9 DXP PERFORMANCE OPTIMIZATION

The main performance issues for the key business-critical pages are depicted in Figure 9-3.

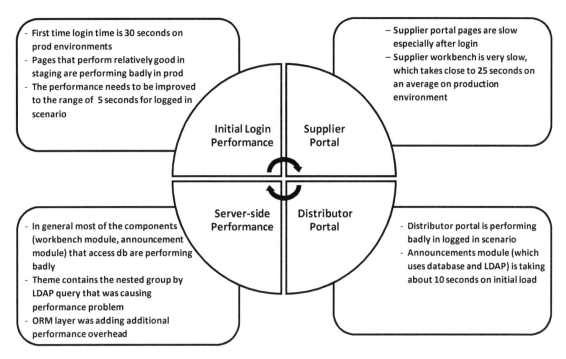

Figure 9-3. Sample performance analysis of business-critical pages

The key performance issues depicted in Figure 9-3 are elaborated as follows:

Initial login performance: The first-time login performance was very bad. This led to bad user experience for the login scenario. Though the pages in the staging environment had good performance, they performed badly in the production environment.

Server side performance issues: The business modules such as workbench module and announcements module had bad performance due to complex LDAP queries involved. Additionally, the object relational mapping (ORM) layer added performance overheads for complex database queries.

Supplier portal and distributor portal: The supplier portal and distributor portal were analyzed to understand the page response time, asset size, asset load time, and other key performance metrics. The server-side components were profiled to understand the performance of all the calls involved in the process.

Following are the key findings from performance analysis and profiling of the components:

- At the presentation layer, the page size or supplier landing page and distributor landing page was more than 2 MB each, leading to higher latency. Each of the landing pages loaded more than 20 JavaScripts and 6 CSS files. There were five images used on the pages.

- Initial login process was very slow due to multiple LDAP and database calls.

- On the server side, the ORM layer was performing complex table joins that further degraded the overall performance.

Further to these, the following specific issues were noticed:

- The database-based modules were querying all records (select * call) from the database. This is done for both "Pending requests" and "Supplier Profiles" tabs.

- This will have the following impact on page performance:
  - Even though we are displaying 20 records (10 each in two tabs), we are querying all records (which would potentially run into thousands).
  - This will increase the data transferred between portal server and database server.
  - This will also increase the initial page load times of the page and the HTML page size.
  - As these queries were done via the ORM tool, it created Java objects for each database record, which could potentially increase the consumed memory.

Recommendations and Improvements

Based on the findings from the performance analysis, the performance engineer recommended and implemented the following suggestions. After applying all the performance optimizations, the page load time was reduced from 120 seconds to 10 seconds.

Presentation Layer Performance Optimizations

Table 9-4 shows the performance optimizations that were implemented at the presentation layer.

Table 9-4. Presentation Layer Performance Optimizations

Performance Improvement	Comments
Merging and minifying CSS and JS files	• Merge all page level CSS (3 in total) and JS (18 in total) files. • JS files are taking the bulk in terms of page size (80% of total page size).
Add all external JS files at the bottom of the page and CSS at top of page	• This will enable the browser to load the Document Object model (DOM) and it is not blocked by JS files. • This will also help to load the CSS fast.
Use CSS sprites and a CDN if feasible. CSS sprites method uses a single combined image and styles to render various image pieces.	• CSS sprites will reduce the number of image requests. • CDNs (like Akamai) will improve the asset response times across geographies.

Server layer performance optimizations

The following performance optimizations were implemented for server side performance optimization:

- The web modules implemented paginated query to only query 20 records, avoiding the full table data loads.

- The query results module implemented pagination that loaded the results on demand using AJAX calls.

- Nested loops (wherein a database query was invoked within a loop resulting in huge number of database calls) containing the database transactions were replaced. A query batching feature was used to avoid frequent database calls.

- A caching framework was used to cache the master lookup values (overall list of suppliers, list of countries, etc.).

- LDAP objects were cached to avoid multiple LDAP queries.

Database Layer Performance Optimization

Following are the performance optimizations done at the database layer:

- Database indexes were created for the columns appearing in the filter conditions of the queries. The table data was partitioned based on the country, as suppliers and distributor data were mainly using country-specific values.

- Denormalized tables (a method in which table structure is flattened out with all needed columns to avoid real-time complex joins) were created to hold the results of complex table joins. These snapshot tables were refreshed every 5 hours. This avoided costly real-time table joins, improving the query performance.

- All the database-specific configurations (such as checkpoint logs, query cache size, sessions, log buffer, recomputation of statistics, etc.) were optimized from a performance standpoint.

- Lookaside database tables and materialized views were created to store the results of frequently used queries.

Chapter Summary

- The main performance best practices for presentation layer performance optimization are: reducing HTTP requests, image optimizations, optimal resource loading, caching, delivery from edge locations, compression, and optimal positioning.

- Key performance metrics are: TTFB, page size, page response time, above the fold time, and perceived response time.

- The main steps in a performance testing framework are: identification of critical transactions, workload analysis, qualitative assessment, quantitative assessment, and prediction.

- The three steps in performance debugging framework are: identification of root cause, performance optimization of the identified component, and verification by testing.

CHAPTER 10

Transforming Legacy Banking Applications to Banking Experience Platforms

Digital transformation is changing the way traditional banks function, and redefining the way traditional banks engage with the customers. Banking organizations want to leverage modern digital technologies to provide engaging customer experiences on all channels—anytime, anywhere. Newer banks are digital native, providing a digital-first banking experience, whereas existing banks are in the process or transforming and digitizing their business processes.

This chapter discusses banking-related digital transformation. We will discuss the key trends in digital banking and various aspects of digital transformation in the banking sector. We propose the key tenets and architecture for a banking experience platform.

This chapter discusses various methods to migrate existing web platforms to DXPs. Though we are discussing topics for banking domain digital transformation methods, reference architecture is applicable for other scenarios as well.

CHAPTER 10 TRANSFORMING LEGACY BANKING APPLICATIONS TO BANKING EXPERIENCE PLATFORMS

Key Tenets of a Banking Experience Platform

Following are the main tenets of a banking experience platform:

- *User-centric experience redesign*: The end user interface should be designed to provide seamless experience across the user journey at all the touch points. The user interface should be personalized based on user preferences, needs, and wants. The user interface design should provide a single-stop view of all activities such as transactions, payments, and deposits.

- *Customer insights gathering*: The banking interfaces should be integrated with analytics software to gather insights about customer activities. Cross-channel analytics (analytics across various channels) helps us to provide more personalized experience and provides targeted content, promotions, and products/services.

- *Optimized business models*: A modern banking experience platform redefines the business processes to provide an automated, frictionless, and seamless user experience at all touch points.

- *Leveraging modern digital technologies*: Next-generation banking experience platforms leverage artificial intelligence (AI) to provide self-learning capabilities; comply with regulations; do form processing, search, and credit scoring; and provide conversational interfaces to provide superior user experience. AI technologies can be leveraged for portfolio planning, wealth planning, chatbots, virtual assistants, and such. Other digital technologies that are applicable in the banking domain are Blockchain (a sophisticated distributed ledger), IoT, big data, APIs, and the cloud.

- *Digital open ecosystem*: Modern digital banks leverage APIs to aggregate all necessary services (such as payment, lending, social, etc.) to create an extensible banking platform. Using microservices and APIs, the next-generation banking platform should be able to aggregate information and integrate with partner systems to provide a frictionless experience.

Attributes of a Next-Generation Digital Bank

Figure 10-1 depicts the core attributes of a next-generation digital bank.

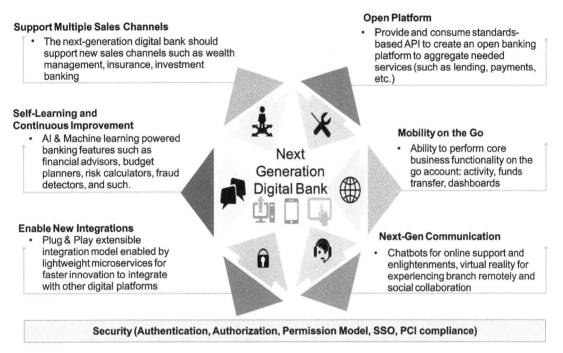

Figure 10-1. Digital bank core attributes

Core attributes of a digital bank are discussed in detail as follows:

- *Support multiple sales channels*: The next-generation banking platform should support various channels such as retail banking, investment banking, wealth management, insurance, and such. Additionally, users should be able to view information and perform transactions seamlessly on any device anytime.

- *Self-learning and continuous improvement*: The DXP should leverage 360-degree insights about customer actions across all channels and use the information to personalize the user experience. AI-driven tools should be used as financial advisors, budget planners, comparators, personalized recommendations, and such that utilize the users' transaction data and other data (such as web analytics data, customer survey data, social data, customer demographics data,

CRM data) for providing relevant recommendations and advice customers in decision making. As part of continuous improvement, the next-generation banking platform should provide automated processes and leverage digital technologies such as IoT, Blockchain, cloud, and robotic process automation (RPA) to continuously improve the user experience, analytics, and business processes. Some examples of innovations using digital technologies are:

- Complete digitization and automation of account opening and customer onboarding process
- Fully paperless forms with e-signature
- AI-powered chatbots, financial advisors, finance planners, and virtual assistants for customer service
- Biometric authentication from a mobile app
- Virtual branch through real-time collaboration
- Branchless digital banking through a mobile app
- Enabling online person-to-person (P2P) payments, digital payments, contactless payment, and e-wallet through a mobile interface

- *Enable new integrations*: The next-generation banking platform should be flexible, to enable new integrations to enable future growth and innovations. The integration model of the platform should be built around a lightweight microservices services model. The lightweight services model plays a crucial role in creating an open banking ecosystem and integrating with other platforms and services.

- *Open platform*: The banking platform should expose the services for third-party applications and services to consume the necessary services. The open platform can be leveraged by bank partners such as payment partners, financial partners (who handle cards, insurance, mutual funds, trading, mortgages, etc.), financial technology (fintechs), merchants, digital partners, lending partners, technology partners, telecommunications companies, etc. The banking platform should provide a marketplace for enabling collaborated development. The marketplace can host the applications, software development kits (SDKs), solutions, widgets, and integrations that can be used by a partner ecosystem.

- *Mobility on the go*: All the main banking functions such as account dashboard, transfers, and such should be available on all devices.

- *Next-gen communication*: The next-generation platform should leverage digital technologies such as chatbots, virtual assistants, virtual reality, and predictive analytics to provide timely alerts, notifications, and promotions, and collaborate with the user. A few examples of innovations in this category are:

  - Robo advisors to recommend in the areas of savings, investment, and portfolios based on users' behavioral data and transaction history

  - Forecasting and self-service tools for wealth management

Security is the key concern that should be enabled across all channels and transactions. Enabling security includes authentication, role-based access, encrypting data, multifactor authentication, secured communications, and compliance to security standards such as Payment Card Industry (PCI).

DXP Features for Next-Generation Digital Bank

Figure 10-2 provides the core attributes of a DXP platform that can be used for next-generation digital banks.

CHAPTER 10 TRANSFORMING LEGACY BANKING APPLICATIONS TO BANKING EXPERIENCE PLATFORMS

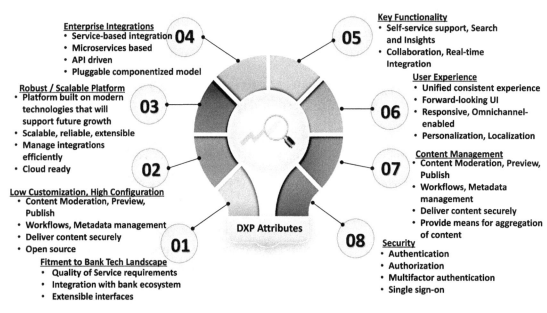

Figure 10-2. DXP Attributes for banking platform

DXPs provide the following features that can be leveraged for building a next-generation banking experience platform. The key DXP attributes as depicted in Figure 10-2 are explained as follows:

- *Enterprise integration*: Support for services, lightweight micro services integration model, and API driven integration can be used by the banking experience platform to expose and consume services for external systems. Robust integration is used for the open banking ecosystem for enhanced collaboration and cocreation of products and services.

- *Robust platform*: The modular design, componentized architecture, and extensible design of a DXP is necessary to provide on-demand scalability for the banking experience platform. Cloud native and cloud support can be leveraged to implement a cloud-first strategy for the banking experience platform.

- *Low customization and high configuration*: DXP packages necessary platform features needed by the banks. Banks can leverage inbuilt DXP features such as workflows, rules engine, content management, personalization, and security to quickly build and deploy the banking experience platform. This can be done mainly through platform configuration with minimal customization.

- *Applicability to bank landscape*: The robustness and scalability of the DXP can be leveraged to meet the quality SLAs of the bank. We could also integrate with the bank's ecosystem with the integration model supported by the DXP. Additionally, the modular architecture of DXP enables the bank to easily extend and add new innovations and capabilities in the future.

- *Functionality*: A DXP provides many of the digital platform features such as search, content management, analytics, collaboration, and such. These can be leveraged to quickly implement the functionality of the banking experience platform.

- *User experience*: DXPs provide forward-looking, responsive, and engaging UIs. A DXP's inbuilt features can be used to build a personalized and omnichannel-enabled user experience.

- *Content management*: A DXP's inbuilt content management features such as content authoring, tagging, editing, archiving, and publishing can be leveraged for creating product content and campaign content for banking customers.

- *Security*: DXPs support multiple security methods such as authentication, authorization, single sign-on (SSO), multifactor authentication, and federated single SSO that can be leveraged to build the banking experience platform.

CHAPTER 10 TRANSFORMING LEGACY BANKING APPLICATIONS TO BANKING EXPERIENCE PLATFORMS

Main Trends in Digital Banking

This section discusses the main technology and business trends in banking.

Technology-Related Trends

The key technology related trends are as follows:

- *Artificial intelligence*: AI technologies can be used in variety of tasks such as rules-based automation, providing personalized recommendations, learning user preferences, providing virtual assistance, customer service, and other functions. AI technologies are preferred for user engagement, cost optimization, and to provide business value differentiation. AI technologies combined with natural language processing methods can provide powerful voice-enabled virtual assistants.

- *Blockchain*: Blockchain manages the distributed ledger, storing the sequence of transactions in a distributed network where data cannot be changed. This enables easier management and tracking of transactions. It can be used for use cases such as fund transfers, title registration, contract registration, and bank to bank transactions for faster execution of financial transactions.

- *Cryptocurrency*: Cryptocurrencies are digital money that is gaining traction for a small percentage of financial transactions. Lack of centralized control and legal restrictions pose challenges in the usage of cryptocurrency.

- *Biometric authentication*: Future authentication mechanisms heavily rely on biometric authentication comprised of fingerprint or face recognition.

- *APIs and microservices*: Microservices-based architecture provides needed scalability, high availability, and reliability for banking platforms. APIs are an essential part of an integration ecosystem.

- *Gamification*: Gamification methods (such as reward points, instant feedbacks, badges, levels, and such) are used to reward loyal customers and establish a deep relationship with the customer. Gamification methods also include incentives, loyalty rewards, and targeted promotions.

- *Virtual reality (VR) and augmented reality (AR)*: These technologies offer immersive experience to end users. In the banking domain, VR technologies can be used for training, process walk-through, end-user demos, and home mortgage support.

- *Internet of Things (IoT)*: Banks can use sensors and other devices at branches and ATMs to obtain insights into customer behavior, and fine tune the services accordingly.

Business Process-Related Trends

The main business process-related trends in banking are as follows:

- *Crowd-based P2P lending*: The crowd platform that enables person-to-person lending

- *Digital-first bank*: A bank that predominantly provides digital channels for core banking functionality with fully digitized bank processes

- *Payment banks*: Digital banks that specialize in digital payments

- *Digital wallets*: Systems for managing e-money

- *Social media banking*: Banking through social media channels (such as Facebook) and messaging channels (such as WhatsApp)

Digital Transformation of Traditional Banks to Digital Banks

This section looks at various options for the digital transformation of traditional banks to modern DXPs.

Reference Technology Architecture for a Digital Bank

A reference technology architecture of a digital bank is depicted in Figure 10-3.

CHAPTER 10 TRANSFORMING LEGACY BANKING APPLICATIONS TO BANKING EXPERIENCE PLATFORMS

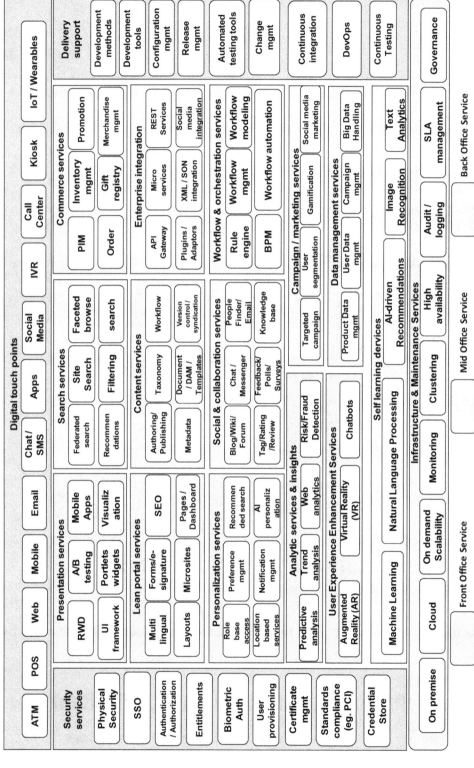

Figure 10-3. Digital bank reference architecture

The next-generation digital banking platform provides modular and services-based modules. As part of front office services we can include lean portal services, presentation services, user experience enhancement services, personalization services, and analytic services. Mid-office services typically include business service modules such as search services, social and collaboration services, workflow and orchestration services, campaigns, self-learning services, and marketing services. The back office includes data management services and content services.

Front Office Services

Presentation service components provide end-user interface and define the overall user experience. This category includes modular UI components such as portlets and widgets, model-view-controller (MVC) UI frameworks (such as Angular, ReactJS), mobile apps, and responsive web design. Presentation service components provide a seamless user experience across all channels and devices and across all user touch points.

Lean portal services play a crucial role in offering an engaging and interactive experience to the end user. Lean portal services include UI components such as forms, dashboards, single-page applications (SPAs) wherein the entire applications consists of a single page, and microsites that provide an interactive experience to the end users.

Personalization services provide contextual and personalized content based on user preferences, user behavior, and context (such as location). The platform should gather user behavioral insights and offer personalized search recommendations based on the insights.

The analytic services and insights module aggregates user actions at all touch points and channels to provide integrated insights that can be used for personalization, prediction, and recommendation. The main categories of analytics are descriptive (that analyze historical data to draw insights), predictive (that look at trends and patterns to predict behavior), and prescriptive (that uses predictive analytics to suggest next steps). Fraud detection can be done based on insights and trends obtained from transaction data. Other uses of analytics are automatic customer segmentation, fraud detection, customer behavior analytics, campaign analytics, sentiment analysis, customer churn prediction, and such.

User experience enhancement services include employing niche digital technologies such as VR, AR, and chatbots for business functions such as training, education, support, promotion, and such. Sensors and IoT devices are also used for real-time service monitoring and prediction.

Mid-Office Services

Mid-office services mainly include enterprise integration services, search services, social and collaboration services, commerce services, self-learning services, and workflow and orchestration services. Integration services form the backbone of the platform extensibility. Integration services provide support for a flexible integration model by supporting various integration methods such as microservices-based integration, APIs, and such. The integration module also supports a pluggable adaptor framework and out-of-the-box integration plugins to social media platforms, ERP systems, and such.

Search services normally include enterprise search engines that index enterprise data and provide relevant search results. Various capabilities in this category are site search, content search, advanced search (faceted search, search filtering, and personalized search). Cognitive search is an emerging technology in this category. Machine learning and natural language processing methods are used in cognitive search to understand natural language queries and provide contextual results by processing structured and unstructured data.

Social and collaboration services provide tools to collaborate and share information. This module includes blog, wiki, forums, communities, chat, messenger, people finder, shared calendar, feedback, surveys/polls, review and rating widgets, and such. Knowledge management systems should be centralized across the enterprise for effective usage of knowledge. Search built on top of a centralized knowledge base can be used for training, e-learning, troubleshooting, and for customer support.

Commerce services provide digital commerce-related functionality such as order management, catalog management, product information management, merchandise management, gift registry, and such. Most of the DXPs provide inbuilt commerce services for quickly developing a digital commerce platform.

Self-learning services include employing AI and machine learning methods to train the model to perform various tasks such as smart recommendations, image recognition, financial advisors, text analytics, and such. AI methods are used to enable chatbots and virtual assistants.

The workflow and orchestration module consists of business process modeling tools, rules engine, and workflow automation tools to model and optimally design business processes.

Campaign/marketing services are used by the sales and marketing team. This module includes capabilities such as customer segmentation, campaign lifecycle management (campaign generation, campaign configuration, campaign execution, and campaign monitoring). Gamification is used for increasing the effectiveness of the campaigns. Social media marketing and social listening are other emerging areas in this category.

Back Office Services

As part of back office services we have data management services that centrally manage customer data, campaign data, and product data in a system of record. With the emergence of big data technologies, organizations need to handle big data to gain insights into customer behavior.

Content services are a quintessential part of experience platforms. This module includes content lifecycle management (content creation, content update, content deletion, content tagging) through business-friendly interfaces. Content workflow management, taxonomy, metadata management, and content version are other capabilities in this category.

Horizontal Services

There are other horizontal services that are included as part of DXPs: security services, infrastructure, and maintenance services.

Security and identity services provide security capabilities such as authentication, authorization, SSO, permission model/entitlements, biometric authentication, user provisioning, certificate management, credential store, and such. Security services ensure that the experience platform is securely accessed by appropriately individuals.

Delivery support involves methods for optimal program management of a digital solution. Modern digital solutions use agile methods to quickly deliver the capabilities to the market. Automation and continuous integration tools are used in release management for agile delivery.

Infrastructure and maintenance services are used in deployment and maintenance of the digital solution. Various capabilities in this module are cloud deployment support, clustering, high-availability support, scalability support, SLA management methods, monitoring infrastructure, and such.

Reference Functional View of Digital Bank

A reference functional architecture of a digital bank is depicted in Figure 10-4. We can use the technical components depicted in Figure 10-3 to realize the functionality.

CHAPTER 10 TRANSFORMING LEGACY BANKING APPLICATIONS TO BANKING EXPERIENCE PLATFORMS

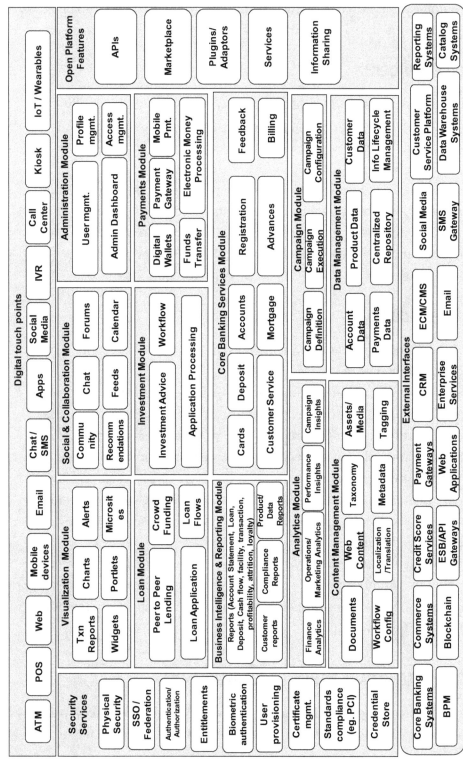

Figure 10-4. Digital bank reference functional view

Figure 10-4 covers the key functional modules of a digital banking experience platform.

As depicted in Figure 10-3, digital touch points provide access channels such as mobile, web, POS e-mail, and others to access underlying digital services.

The visualization module provides various charts, reports, widgets, and portlets for end users and administrators. The visual modules are configurable so that users can customize the presented data based on filter and configuration values.

The social and collaboration module provides various collaboration features such as communities, chat, forums, feeds, and calendar needed for banking experience platforms.

The administration module enables administrators to manage user accounts, workflows, business rules, business processes, and user access. Normally, the administration module consists of an admin dashboard to provide a unified view of all admin functions.

The loan module provides features for lending, such as crowd funding, peer-to-peer lending, application processing, workflow handling, and such.

The investment module includes investment application processing, investment advice, and reports.

The payment module includes digital wallets, funds transfer, mobile payment, and payment gateway integration.

The banking services module includes key banking services such as account handling (current account, savings account), card handling (debit and credit card), deposit handling (term deposit), user registration, feedback handling, mortgage, HR, billing, and customer service.

The business intelligence (BI) and reports module handles various functions such as reports processing (report creation, reports configuration, report generation, reports delivery) for various banking functions, compliance, customer insights, and products/transactions.

The analytics module has features to provide insights about customer behavior, transactions, operations, campaigns, and performance.

As part of the campaign module, the administrators should be able to define/create the campaign for specific channels, events, and products. Campaign administrators should be able to define the triggering event and deliver the campaigns based on customer insights (offering products that a customer is most likely to buy). Campaign administrators should monitor campaign responses and effectiveness and use the information to fine tune the campaign strategy.

The content management module provides various features to manage and deliver content. These include features for authoring content, editing content, tagging content, and such. Other content-related functions such as workflow management, taxonomy management, metadata management, content translation, document management, digital asset, and media management are part of this module.

The data management module provides centralized management of account data, product data, transaction data, customer data, and payments data in a centralized repository.

Normally, banking systems need to interface with various systems such as CMS, search, credit score services, CRM, social media, reporting systems, BPM (for business process management), Blockchain systems, enterprise web applications, e-mail, SMS gateway, catalog systems, data warehouse systems, and such.

Security services include enterprise security-related functions such as physical security (perimeter security), SSO, authentication and authorization, biometric authentication, user provisioning, certificate management, standards compliance, and credential store. We discussed this as part of Figure 10-3.

Open platform features include functionality that extends and evolves the digital applications. This includes services that expose the core platform functionality, such as APIs and a marketplace that hosts various libraries and applications and information sharing modules.

Technology Transformation

The technology transformation from traditional web technologies to integrated DXPs is depicted in Figure 10-5.

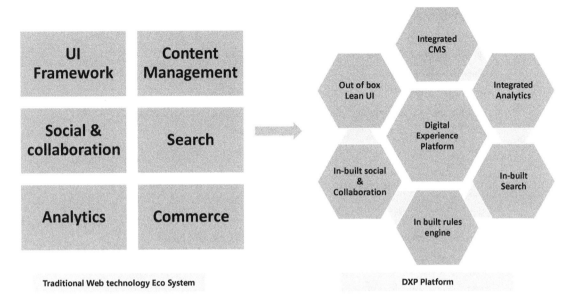

Figure 10-5. Technology transformation

As depicted in Figure 10-5, traditional web technologies need to be integrated with various systems such as CMS, search, commerce, workflow, and campaign to create various needed capabilities. However, a DXP provides all the needed capabilities in an integrated way to provide a seamless experience to the end user and provide 360-degree insights to the business.

The impact of migration from traditional web technologies on user experience, customer insights, and business agility is elaborated in Figure 10-6.

Traditional Web Technologies	Integrated DXP
User Experience	
Heavy weight UI components	Light weight lean UI
Heavily server based	Client side aggregation
Desktop-first design	Mobile-first design
Disjoint and inconsistent UI	Uniform dashboard experience
Analytics and insights	
Minimal or absence of cross channel analytics	In built cross channel analytics
Minimal personalization feature	Information and services personalized based on insights.
Business Agility	
Longer deployment cycles	Shorter deployment cycles
Requires integration with multiple external tools.	In built automation, self-service and business productivity improvement
Challenges in absorbing changes	Responsive to change

Figure 10-6. Impact of migration to DXP

Main Digital Transformation Methods

There are broadly two main methods for the digital transformation of traditional banks:

- *Digitizing existing banking systems*: We can adopt this option for large traditional banks that have heavily invested in legacy platforms. We will enable the key digital capabilities on top of existing banking systems. This helps banks to leverage existing investments.

- *Reimagining the banking experience*: In this bottom-up option we will completely revamp the existing access channels, user experience, business processes, and services to create a digital native banking experience platform. This is suitable for banks that want to start fresh and the ones that have minimal investments in legacy banking systems.

Digitizing Existing Banking Systems

Let's look at the main digital enablers for this option, which can be provided by reengineering existing functions:

- *Providing omnichannel capabilities*: Banks can provide omnichannel capabilities through mobile apps or mobile web so that banking services can be accessed from all channels.

- *Digitizing banking business models*: Existing banking processes can be digitized by using modern digital technologies. For instance, business process flows can be digitized using BPM tools, providing faster and more reliable execution of business processes. Other business processes such as online registration, loan application, information sharing, customer support, and incident management can be digitized using modern digital technologies such as online forms, workflow management, collaboration software, and incident management systems. Systems of engagement should be fully integrated with systems of record.

- *Digital facelift for user experience*: The end user experience (mainly for retail banking) can be reimagined by redesigning the user interface. We can adopt lean architecture providing dashboard experience, personalization, search, and other intuitive information discovery and navigation features.

- *Service enablement*: Banks can build microservices on top of existing APIs or services that are invoked from the user interfaces. If there are no prior services, we need to identify the logical grouping of legacy functions that can be transformed into microservices.

- *Automation*: Identify repetitive jobs and redundant steps that can be automated. Identify human-intensive processes and explore opportunities for automating the steps. Leverage RPA to automate rule-based tasks, data entry tasks, and repetitive tasks to reduce process time and minimize manual errors.

- *Two-speed digital enablement*: This model provides an engaging user experience for end users with rapid innovation. This involves providing granular microservices needed for the fast changing UIs that interface with stable, less frequently changing, back-end web services.

- *Standardization and centralization*: Standardize processes and technologies across the banking ecosystem. Remove unnecessary and redundant or duplicate process steps and consolidate processes.

Create a centralized process management and centralized data management through a system of record. Eliminate information and process silos through centralization.

Various phases of digitization of the banking processes are depicted in Figure 10-7.

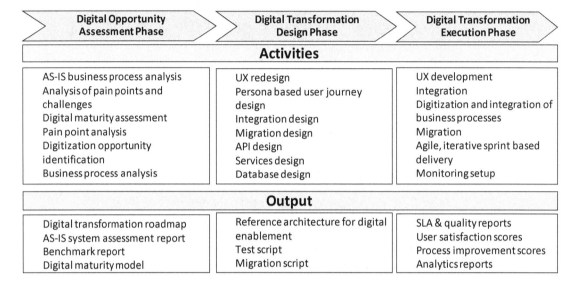

Figure 10-7. Phases in reimagining banking processes

Various phases depicted in Figure 10-7 are discussed as follows:

Digital Opportunity Assessment Phase

In this phase, we need to do a thorough as-is analysis of existing systems and business processes. During as-is analysis we will look at the following:

- Analysis of existing technical ecosystem, standards, and technologies
- Creating an inventory of existing functional modules
- Analysis of existing business processes

During the process we will identify the existing pain points and challenges. We need to interview the key stakeholders and end users to understand the pain points. The typical pain points and challenges are given as follows:

- User interfaces not compatible with mobile devices
- High page response times and performance issues for key banking transactions

CHAPTER 10 TRANSFORMING LEGACY BANKING APPLICATIONS TO BANKING EXPERIENCE PLATFORMS

- Longer process times for core business processes such as user registration, account activation, funds transfer, and such.

- Challenges in onboarding new functionality

- Lot of manual effort spent on repetitive tasks (such as checking for credit scores, application form completeness, and such)

- Longer deployment times for new releases or incremental releases

As part of the digital maturity assessment, we assess the current systems, processes, and culture. The digital maturity assessment can be used to define the digital road map based on their current maturity levels. Listed in Table 10-1 are key points that will be assessed during the digital maturity assessment. Organizations can leverage the points given in Table 10-1 to achieve leading-level digital maturity.

Table 10-1. Key points in Digital Maturity Assessment

Category	Beginner-Level Digital Maturity	Moderate-Level Digital Maturity	Leading-Level Digital Maturity
Organization culture	• Lack or minimal innovative culture. • Minimal usage of collaborative tools and social channels. • Lack of incentive for collaboration • Risk averse and resistant to change • Data managed in silos • Processes and software are not customer-centric.	• Organization provides limited support for collaboration and innovation. • Centralized customer data management	• Customer-centric products and services, and the organization continuously aims to improve the customer engagement based on feedback. • Gamified applications and incentives for collaboration for improved productivity • Employees are encouraged to experiment with tools and technologies to improve customer engagement. • Cocreation of products and services with involvement of partners, end users, and business stakeholders. Data management tools are leveraged to manage quality of centralized data. Organization continuously innovates to meet customers' expectations and needs.

(continued)

Table 10-1. (*continued*)

Category	Beginner-Level Digital Maturity	Moderate-Level Digital Maturity	Leading-Level Digital Maturity
Business process	• Existing processes involve heavy manual and human intensive tasks • Siloed processes per each department and channel • Processes are not user friendly. • Business processes are not assessed on a continuous basis. • Absence of process monitoring and governance tools • Absence of straight-through processing	• Business processes are partially automated.	• New processes are digital native. • Existing processes are revamped/reengineered to become compatible with modern digital systems. • All repetitive and mundane work is fully automated using digital technologies. • Integrated business processes (such as campaign management, customer services, sales) across all channels and departments • User-centric processes • All business processes are continuously assessed and monitored, governed, and the feedback is used to improve the processes. • Straight-through processing • Lean and agile processes to improve time to market

(*continued*)

Table 10-1. (*continued*)

Category	Beginner-Level Digital Maturity	Moderate-Level Digital Maturity	Leading-Level Digital Maturity
Leadership	• Digital strategy not defined • Leadership has not fully understood the potential of digital technologies and digital value proposition. • Minimal support and sponsorship for digital programs	• Partially defined digital strategy applicable only for specific initiatives and departments	• Well-defined and fully supported digital strategy with tracking metrics and KPIs. Main KPIs are customer lifetime value, repeat transaction rate, customer satisfaction index, customer engagement score, customer wallet share, employee engagement score, and growth rate. • Digital value proposition is fully defined and tied to digital technologies and overall digital strategy. • Heavy usage of data-driven decision making with the help of digital technologies • Clear strategy for build, buy, or collaborate for developing needed capabilities
Governance	• Absence or minimal monitoring and governing processes • Roles and responsibilities to implement digital strategy not defined. • Processes for change management, business continuity not defined. • Business processes not fully documented.	• Governance processes are defined for specific programs. • Limited documentation of governance processes and policies	• Well-defined roles and responsibilities for implementing digital strategy • Fully defined processes for change management and business continuity • Monitoring infrastructure to assess the effectiveness of digital strategy • All policies, processes are fully documented in a centralized knowledge base and used for self-learning and training.

(*continued*)

Table 10-1. (*continued*)

Category	Beginner-Level Digital Maturity	Moderate-Level Digital Maturity	Leading-Level Digital Maturity
User engagement	• Users' needs and wants not fully understood. • Absence or minimal self-service features	• Moderate support for personalization and user preferences management	• Digital channels are fully focused on users' needs and wants through persona-based experience. • The organization proactively anticipates customer requirements and leverages emerging digital technologies to provide a seamless user experience. • Proactive user engagement across all channels including social media • Cross-channel frictionless processes • Digital channels provide self-service and decision-making tools. • Cross-channel analytics is used to understand customer behavior and use it for personalization and prediction.
Collaboration and Social media interaction	• Absence of social marketing • Minimal or absence of collaboration	• Partially integrated social channels • Partial usage of collaborative tools such as chat, forums, groups, communities	• Fully integrated social analytics and support for social marketing • Fully integrated cross-channel collaboration across sales, marketing channels • Collaboration with all internal departments to align with digital vision • Open banking ecosystem to provide a platform for partners to collaborate and integrate

(*continued*)

Table 10-1. (*continued*)

Category	Beginner-Level Digital Maturity	Moderate-Level Digital Maturity	Leading-Level Digital Maturity
IT alignment	• IT team is not fully aligned with organization's vision. • IT team heavily uses legacy technologies for new products and services. • Absence of efforts to modernize or service-enable existing eco system.	• IT team occasionally does product evaluation to assess the right fitment of products.	• IT team is fully aligned with digital strategy. • Most of the new services and products are available as service on cloud. • IT team ensures that modern and innovative digital technologies (such as customer experience tools, AI, big data tools, cognitive computing, advanced analytics, IoT) are fully leveraged for customer engagement. • IT strategy constantly looks to automate existing business processes (such as workflows, document processing, compliance checks, and leverage the power of AI for deeper customer engagement. • IT team continuously evaluates emerging technologies to implement digital strategy.
Business agility	• Longer time to market. Typical release cycles last from 6–8 months. • Business rarely gets feedback from end users and does	• Business obtains end user feedback very rarely.	• Well-defined and agile processes that are responsive to business needs and expectations • Continuous competitive benchmarking and continuous improvement • Faster time to market • Quick to integrate and implement innovative technologies • Typical releases happen on a monthly basis using continuous integration tools.

(*continued*)

Table 10-1. (*continued*)

Category	Beginner-Level Digital Maturity	Moderate-Level Digital Maturity	Leading-Level Digital Maturity
Analytics	• Absence of analytic tools. • Manual reporting of traffic reports	• Support for analytics tools for specific channels • Basic analytics reporting and dashboards	• Fully integrated cross-channel analytics to provide 360-degree insight about customer activities. • Support for predictive analytics • Insights-driven personalization, and recommendation and financial advisory
Infrastructure	• On-premise deployment model	• Mix of on-premise and cloud deployment model	• Fully cloud enabled providing on-demand scalability
Data management	• Data distributed across multiple systems • Absence of data management tools • Unstructured data (text, e-mail, phone, chat, videos, blog post) • Duplicate and redundant data	• Data consolidation done partially • System of record exists partially	• Centralized data management through system of record (SOR) for data segments such as customer data, transaction data, payment data, deposit data, etc. • Data quality tools are used for data management. • Structured and unstructured data is processed for advanced analytics and used to get a single view of the customer.

As part of the digital maturity assessment, we also identify the digital opportunities. The digital opportunities are categorized and prioritized based on their business impact and end user impact. Table 10-2 provides insights into available opportunities.

Table 10-2. *Categorized Digital Oppurtunities*

Category	Digital Opportunity	Business Impact	Customer Impact	Overall Priority
User experience	Make the UI mobile enabled	High	High	High
	Provide customer and admin dashboard	High	High	High
	Provide prefilled minimal forms for faster registration	High	High	High
	Reduce the steps in approval workflows and automate workflow steps wherever possible	High	High	High
Sales channels	Enable other channels such as investment banking, wealth management, commercial banking.	High	Medium	High
	Provide self-service to enable sales personnel	High	High	High
Customer service	Provide niche capabilities such as chatbot, VR, AR	Medium	High	High
	Employ AI technologies to provide personalized recommendations and financial advising capability	Medium	Medium	medium
Customer engagement	Provide self-help decision-making tools such as financial planners	High	High	High
	Provide faster and frictionless processes: provide 1-click account opening, 1-click loan approval, minimal field registration form with 1-click registration	High	High	High

At the end of the assessment phase, we create a digital transformation road map, as-is system assessment report, competitive benchmark report, and digital maturity model.

Digital Transformation Design Phase

In this phase we mainly design the key digital transformation elements such as user experience design, migration, integration, API design, services, database, and such.

We have extensively discussed the UI design as part of Chapter 4 and integration design as part of Chapter 5.

At the end of this phase, we will have the various design and reference architecture documents.

Digital Transformation Execution Phase

During this phase, we do the actual activities designed in the previous phase. We develop/redesign the user experience, redesign the integration model, and migrate the necessary code and data. All the planned business processes will be digitized and automated wherever possible. The capabilities are delivered in shorter and iterative sprints for quicker time to market. Once all the necessary capabilities are set up, we can set up the monitoring infrastructure to monitor the SLAs. We will track the KPIs such as user satisfaction scores, process improvement scores, SLAs, and quality reports.

Digital Transformation Road Map

A sample digital transformation road map is depicted in Figure 10-8. Organizations can use this as a reference for defining a digital transformation road map.

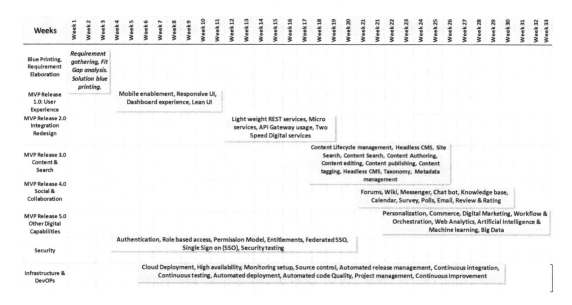

Figure 10-8. Digital transformation road map

Reimagining the Digital Banking Experience

This section defines all the tools, methods, and processes needed for redefining the banking experience.

CHAPTER 10　TRANSFORMING LEGACY BANKING APPLICATIONS TO BANKING EXPERIENCE PLATFORMS

Digital Transformation Tools and Methods

Table 10-3 provides sample tools and methods that can be used for digital transformation.

> **Note** The products and tools mentioned in the table are just a sample list and are provided only for education purposes. It is not an exhaustive list or a recommendation. All product names, trademarks, logos, and brands are property of their respective owners.

Table 10-3. Sample Tools and Methods for Digital Transformation

Digital Experience Capability	Key Features	Key Tools for Migration	Key Methods for Migration
User experience	Mobile enablement, responsive UI, dashboard experience, lean web-oriented architecture, mobile app, forms, microsites, multilingual, layouts, easy and fast information discovery	• Cordova for cross-platform development • Robotium and Selenium for test automation • Robolectric and Mockito for mobile app unit testing • Bootstrap, CSS3-based responsive design • Cognitive search tools	• Use MVC UI frameworks (Angular, ReactJS) for lightweight widgets and personalized dashboard. • Develop existing dynamic contents/pages with UI JavaScript frameworks and static contents/pages with headless CMS. • Convert existing static JSP/HTML into responsive, using CSS3 Media queries. • Web-oriented architecture • Information architecture redesign • Search implementation

(*continued*)

Table 10-3. (*continued*)

Digital Experience Capability	Key Features	Key Tools for Migration	Key Methods for Migration
Integration model	Lightweight REST services, microservices, API gateway usage, two-speed digital services	• MuleSoft, IBM ESB, Micro services, IBM BPM, Jboss BPM, TIBCO, MQ, Apache Kafka, Apache Camel, ServiceMix, WSO2, Spring Boot	• Service enable existing interfaces • Develop granular microservices on top of existing legacy services to implement 2-speed digital services. • Deploy microservices in containers for individual scalability.
Social and collaboration	Forums, wiki, messenger, chatbot, knowledge base, calendar, survey, polls, e-mail, review, and rating	• Liferay SocialOffice, MS SharePoint, Skype, Adobe Connect, IBM Connections, Zoho Connect, Google G-Suite, Yammer, Jive, OpenText FirstClass Collaboration Suite, Slack, OneDrive,	• Implement centralized knowledge base and enable search on knowledge base. • Implement collaboration capabilities using tools. • Harness collective intelligence using forums and communities. • Integrate external social platforms for enhanced user engagement.

(*continued*)

Table 10-3. (*continued*)

Digital Experience Capability	Key Features	Key Tools for Migration	Key Methods for Migration
DevOps	Source control, automated release management, continuous integration, continuous testing, automated deployment, automated code quality, project management, continuous improvement	• Key CI tools: Jenkins, Ansible, Hudson, Puppet, Chef, Bamboo • Build Tools: Maven, ANT, Gradle • Source control: Git, Bitbucket • Code Quality: SonarQube, CheckStyle, Appscan, PMD • Testing: SOAPUI, Junit, Jmeter, Nunit, Corbertura, Fortify, Selenium • Containers: Docker, Kubernetes • Project Management: Jira	• Implement continuous integration using CI tools. • Automate release management pipeline using automated tools. • Set up notification for build and quality reports.
Web analytics	Track user behavior actions to get insights, cloud-based reports, performance monitoring, traffic reports, exit reports.	• Google analytics, open web analytics, Piwik, Adobe marketing cloud, IBM Unica, LiveChat, WebTrends	• Include the necessary JavaScripts to the page. • Populate the JavaScript variables with the runtime values.

(*continued*)

Table 10-3. (*continued*)

Digital Experience Capability	Key Features	Key Tools for Migration	Key Methods for Migration
Content management	Content authoring, content editing, content publishing, content tagging, headless CMS, taxonomy, metadata management	• Drupal, Wordpress, Joomla, Alfresco, LiferayCMS, Kentico, Adobe AEM	• Migration of contents from file system, DB to content management systems • Create reusable content layouts and structures. • Create metadata strategy for content tagging and easier information discovery. • Provide content services to implement headless CMS.
Other digital experience capabilities	Search, personalization, commerce, digital marketing, workflow and orchestration	• Search: Elasticsearch, Solr, Lucene, Splunk, Jena • Digital Marketing: OpenEMM, CampaignChain, IBM Unica, Oracle Eloqua • Personalization: Adobe Target, Google Optimize 360, HubSpot, Marketo • Workflow: Activiti, Jboss JBPM, Copper, Camuda	• Enable site search, enterprise search using search tools. • Implement role-based access and targeted content delivery using personalization. • Enable commerce features using commerce plugin. • Promote campaigns using digital marketing. • Implement business processes using workflow and orchestration tools.

(*continued*)

Table 10-3. (*continued*)

Digital Experience Capability	Key Features	Key Tools for Migration	Key Methods for Migration
Security	Authentication, role-based access, permission model, entitlements, federated SSO, single sign-on (SSO), security testing	• SSO: Okta, OpenSSO • Authentication: CAS, OpenAM • Security testing: OWASP Zed Attack Proxy (ZAP) • Standards: Oauth, OpenID, SAML	• Service enable existing interfaces. • Develop granular microservices on top of existing legacy services to implement 2-speed digital services. • Deploy microservices in containers for individual scalability.
Artificial intelligence and machine learning	Self-learning, continuous improvement, text analytics, predictive analytics, chatbots, virtual assistants, intelligent recommendation engines, robo advisors, process automation	• NLP: OpenNLP • Key Tools: H2O.ai, Apache PredictionIO, IBM Watson, Google TensorFlow • API.ai, Facebook messenger platform, Botsify, Telegram bots, Botkit, ChattyPeople	• Train the models using machine learning algorithms. • Leverage AI and ML tools for implementing the recommendations, search and chatbots.
Big data	Structured and unstructured data processing, real-time insights	• Big data processing: Apache Spark, Apache Hadoop • NoSQL DB: Apache Cassandra, MongoDB, CouchDB, • Search: Splunk	• Implement map reduce framework to process big data. • Implement big data to process structured and unstructured data (text, e-mail, video, etc.) processing to get 360-degree insights.

Chapter Summary

- The main tenets of the banking experience platform are user-centric experience redesign, customer insights gathering, optimized business models, leveraging modern digital technologies and digital open ecosystem.

- Core attributes of a digital bank are support for multiple channels, self-learning and continuous improvement, enabling of new integrations, open platform, mobility on the go, and next-gen communication.

- The key DXP features that can be used for building the banking experience platform are enterprise integration, robust platform, low customization and high configuration, fitment to bank landscape, functionality, user experience, content management, and security.

- The main technology trends in banking are AI, Blockchain, cryptocurrency, APIs and microservices, Gamification, VR and AR, and IoT.

- The main business trends in banking are crowd-based P2P lending, digital-first bank, payment banks, digital wallets, and social media banking.

- Technical reference architecture for the banking experience platform consists of modules such as lean portal services, presentation services, user experience enhancement services, personalization services and analytic services, search services, social and collaboration services, workflow and orchestration services, campaign, self-learning services and marketing services, data management services and content services, security services, delivery support services, and infrastructure and maintenance services.

- Functional reference architecture for the banking platform consists of visualization module, social and collaboration module, administration module, loan module, investment module, core banking services module, business intelligence (BI) and reports module, analytics module, campaign module, content management module, and Data management module.

CHAPTER 10   TRANSFORMING LEGACY BANKING APPLICATIONS TO BANKING EXPERIENCE PLATFORMS

- Main methods for the digital transformation of traditional banks are digitizing existing banking systems and reimagining the banking experience.

- Digitizing existing banking systems is enabled by methods such as providing omnichannel capabilities, digitizing banking business models, digital facelift for user experience, service enablement, automation, two-speed digital enablement, and standardization and centralization.

- Reimagining the digital banking experience includes phases such as digital opportunity assessment phase and data management.

- Key areas that will be assessed during the digital maturity assessment are organization culture, business process, leadership, governance, user engagement, collaboration and social media interaction, IT alignment, business agility, analytics, infrastructure, and data management.

PART V

End to End Case Study

CHAPTER 11

End to End DXP Case Study

This chapter discusses an end to end DXP case study. We have chosen a business to business (B2B) scenario to showcase how digital experience platforms can be used for all scenarios. In this case study, we discuss the detailed requirements, DXP solution fitment, and implementation.

We have considered a vehicle dealer management solution for this B2B case study.

Drivers and Key Requirements of the Dealer Platform Case Study

Given as follows are the main drivers for the next-generation dealer management digital platform:

Next-generation user experience that engages customers and reduces complexity:

- The digital platform should provide a consistent, omnichannel and personalized user experience across all touchpoints.
- The digital platform should provide self-service tools for dealers, sales executives, and support personnel.
- The user experience should be accessible on any device anytime.
- Identify and improve customer engagement KPIs.
- Provide collaboration across dealers, customer, sales, and marketing team.

Business process improvements:

- The digital platform should reduce the overall time needed to gather all needed information.
- Seamlessly integrate the platform with various back-end systems and services.
- Improve sales cycle time.
- Improve sales process and service tracking processes.

Improved information management:

- Improve the quality of lead/opportunity data and revenue recognition data.
- Improve integration with the order and service management system.
- Optimize business processes related to sales, service, marketing, and CRM.

Improved customer satisfaction:

- Improve the customer experience by providing a seamless sales and service experience across all channels.
- Improve the dealer experience and productivity.

Scalable and robust platform:

- Develop a modern platform capable of scaling to future demands and scale for increased dealer operations.

Architecting the Next-Generation Dealer platform

This section looks at various phases of architecting the next-generation dealer digital platform.

Pain Point Analysis in Current Systems and Processes

The first step in the digital transformation of the dealer platform is to identify all the challenges and pain points with the existing system, as detailed in Table 11-1.

Table 11-1. *Pain Point Analysis*

User Persona	Pain Points
Dealer	• Dealers have to repeat entire processes when making multiple changes. • Inability to accept electronic signatures • There are too many unneeded applications. • Dealers can't identify priority messages. • Application flow does not follow process flow used within the dealership. • Common UI standards are not followed. • No UI standards are followed. UI is not consistent for internally created applications. • Lack of standards for vendor sites • No ability to "alert" dealers when action is needed • Navigation menus and information architecture are not intuitive. • Absence of contextual search feature
Administrators	• Applications lack a robust, real-time, error checking/validation feature. • Lot of manual processes and need to go to multiple applications to complete a single business process • Inability to easily upload photos where needed/helpful • No ability to customize application placement • No formal process to remove/update information • Not able to identify priority messages • Information is not automatically populated. • Look and feel is not up to date.
Customers	• The existing system takes lot of time in completing the processes. • There is a lot of manual process involved, which seems outdated. • Too much of redundant information is being asked • Lack of personalization and contextual information • Difficulty in finding relevant information quickly. • Users need to access multiple applications to complete a single business process.

CHAPTER 11 END TO END DXP CASE STUDY

Solution Tenets of Next-Generation Dealer Platform

Based on the pain point analysis, the key solution tenets of the next-generation dealer platform were developed, as listed in Table 11-2.

Table 11-2. Solution Tenets

Solution Tenet	Solution Description
A responsive site omnichannel-enabled platform	• Responsive application that can be accessed from desktop or any handheld device • Multilanguage feature for localization
Improve overall efficiency by streamlining processes and incorporating new technologies	• Application workflow mapped to business processes • Consistent customer information throughout original equipment manufacturer (OEM) systems and dealer CRMs • Single sign-on to access enterprise wide applications • Application portfolio rationalization that will obsolete applications supporting redundant processes and are duplicating some processes • Real-time business and data validations
Enhance communication by updating and expanding delivery options such as redesigned messaging, e-mail, chat, user level alerts and notifications.	• Support for chat, blog, communities, wiki, alert / notification requirements • Two-way communication channel with online help, chat functionality
Improve platform availability.	• Application available even during off business hours • Disaster recovery strategy that predicts recovery points and recovery through secondary sites • Load balancing strategies to distribute transaction loads to multiple servers

(continued)

Table 11-2. (*continued*)

Solution Tenet	Solution Description
Improve navigation throughout the system.	• Intuitive UI/UX experience • Personalized experience like customizable views, customizable dashboard, and app store-like features
Provide powerful search capabilities	• Provide contextual enterprise search capabilities. • Fast and easy retrieval of relevant information
Integrate content management	• Consistent and seamless information available to customer throughout OEM, dealers, and third-party sales channels
Collect usage analytics to enable continuous improvement.	• Dealer sales performance and customer satisfaction index monitoring • Dealer surveys on new product and application features
Utilize best practices for architecture, design, development, and testing.	• Standardized integration formats and UI guidelines with vendor sites • Scalable and efficient application platform
Reduce custom code	• Applications portfolio rationalization • Capability to add new contents and functionalities
Extensibility for new business models launch and support	• Enable workflow for mobile devices • Telematics-based subscription offers
Seamless extension for financing and leasing options	• Online credit application • Online payment estimator for finance and leasing

CHAPTER 11   END TO END DXP CASE STUDY

The solution tenets and features provided by the next-generation dealer platform are depicted in Figure 11-1.

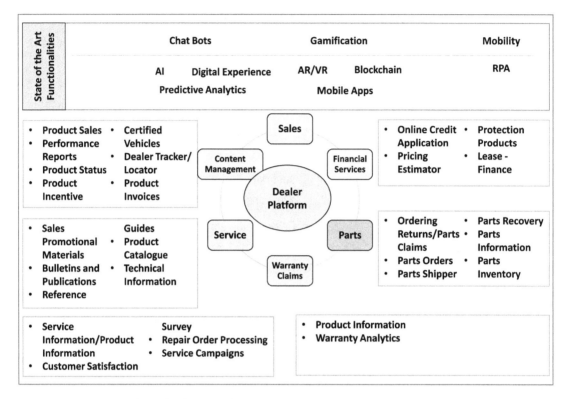

Figure 11-1. *Dealer platform solution tenets and features*

Solution Design Principles

The key solution design principles are given in Table 11-3.

Table 11-3. Solution Design Principles

Solution Design Principle	Core Implementation Points
Continuous availability	• Set up monitoring infrastructure and real-time alert mechanism. • Conduct scalability and availability testing. • Implement multi-node cluster for failover for in house deployment. • Leverage cloud deployment option. • Set up disaster recovery environment. • Implement failover mechanism.
Proven standards and industry best practices	• Leverage open standards for development for future extensibility. • Adopt industry best practices for performance and security. • Create checklists and review gating criteria to ensure that standards and best practices are followed.
Compliance to legal regulations	• Compile all the regulatory and legal policy related requirements (such as security related regulations, data retention-related policies and such) and ensure testing for the same.
Modular and flexible design	• Choose standards and open source technologies and products wherever possible to avoid vendor lock in. • Implement layered and loosely coupled architecture, with each layer having distinct responsibility. • Create modular solution components that can be easily extended.
High usability	• Implement accessibility standards. • Conduct usability testing on all supported browsers and devices. • Provide contextual help, FAQ, and contextual menus. • Provide intuitive information architecture and user-friendly navigation structure. • Provide consistent user interface. • Provide search for all interfaces.
Service-oriented architecture	• Use services for exposing and consuming services. • Implement lightweight REST-based services to integrate with external interfaces.

(continued)

CHAPTER 11 END TO END DXP CASE STUDY

Table 11-3. (*continued*)

Solution Design Principle	Core Implementation Points
Quality attributes	• Get the signed-off SLAs for quality attributes such as scalability, availability, performance. • Conduct iterative testing for verifying all the signed-off SLAs. • Set up monitoring infrastructure to monitor the SLAs in real time.
Security	• Implement layer-wise security. • Create a security check list and use it for reviews. • Conduct iterative security testing.
Lean web-oriented architecture	• Develop lean UI interfaces based on latest JavaScript frameworks. • Provide easy to use UI dashboards to provide an integrated unified view of all information. • Build lightweight UI widgets instead of heavier server-side components.

An overall next-generation dealer platform with various features is depicted in Figure 11-2.

CHAPTER 11　END TO END DXP CASE STUDY

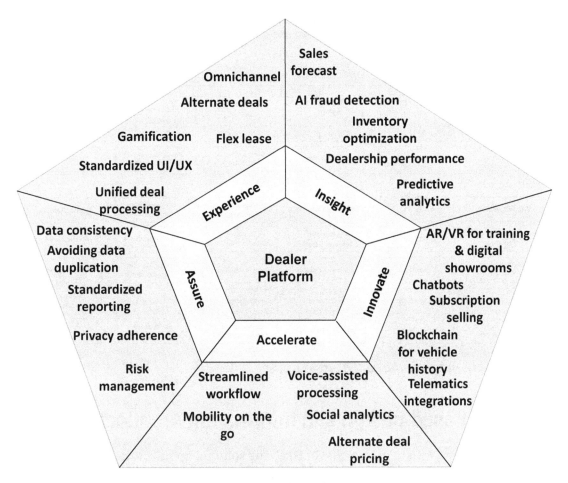

Figure 11-2. *Next-generation dealer platform features*

Persona-Based Information Architecture

In order to improve information discoverability and provide an easy to use and easy to find experience, the new dealer platform provides persona-based features. The user interface provides the dashboards covering the needs and wants for the user persona. The persona-based feature mapping is depicted in Figure 11-3.

307

CHAPTER 11   END TO END DXP CASE STUDY

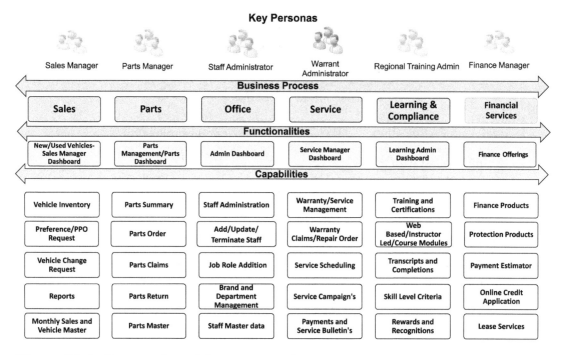

Figure 11-3. *Persona-based feature mapping*

Persona-Based Design and Information Architecture

User centricity is the main design goal of DXP. The solution architecture, UI design, and information architecture are all designed with the end user in mind. "Persona-based design" is one of the key elements of UI design. A persona represents a group of users with similar needs, similar behavior, and similar information goals with similar navigation patterns. We will identify all the user personas for a given digital solution and this will help us to identify the UI design, information architecture, and DXP capabilities that need to be leveraged. In Table 11-4 we have identified a few user personas and mapped them to the DXP capabilities.

Table 11-4. User Persona and DXP Capabilities

Information Goal	Daily Tasks and Navigation Behavior	DXP capabilities required
Sales Manager Persona		
The sales manager is looking to have a single-stop shop for all the vehicle sales reports, vehicle change request reports, and vehicle inventory information and would like to use the same to manage the digital sales.	The user starts using the online sales dashboard provided to get access to the vehicle inventory and other vehicle-related sales information. The user can create a custom, personalized dashboard that provides instant access to all sales information. The user also changes the look and feel and to personalize preferences. The user wants to find the information quickly with minimal page hops.	The **personalized dashboard**, accessible through multiple channels like mobile or desktop, provides the flexibility to have access to a readily defined view with only the banking functionalities that the user uses regularly, making it simple for the user to track. **Custom, personalized alerts** provide real-time updates to the user on all financial transactions. Provide **omnichannel and responsive UX** to manage transactions anytime, anywhere, and effectively manage the finances. DXP should provide a personalized user interface with contexual search capabilities and self-service features.
Warrant Administrator Persona		
The warranty administrator handles multiple operation areas such as service management, claims, repair, orders, campaign and payment and currently has to scan through multiple applications to understand the status of the daily operations.	The user wants an integrated platform. The user wants real-time and configurable reports for claims, repairs, orders, and campaign and payment information.	DXP should be configured to provide an integrated dashboard view that can pull the data from desperate systems and provide the user an end-to-end view of all operations and also provide the user with a personalized interface that can be modified to suite his/her needs. DXP should provide reporting capability that can be configured/customized to effectively monitor operations. DXP should provide a customizable view by adding/removing UI widgets.

(*continued*)

Table 11-4. (continued)

Information Goal	Daily Tasks and Navigation Behavior	DXP capabilities required
Regional Training Admin Persona		
Regional training admin wants to reuse the organizational internal knowledge base to provide user training and certification. The regional training admin wants to use the centralized knowledge base for web-based training. The regional training admin carries out a rewards and recognition program.	The user uses the centralized content management system to access the training content and course content. The user leverages the gamification feature of the platform to promote learning.	DXP should provide real-time collaboration features such as chat, blog, wiki, messenger, and information sharing platform. The content management feature of DXP should be used to create a knowledge repository and provide a search feature for efficient information discovery.

Functional View of the Next-Generation Dealer Platform

A functional view of the next-generation dealer platform is depicted in Figure 11-4.

CHAPTER 11　END TO END DXP CASE STUDY

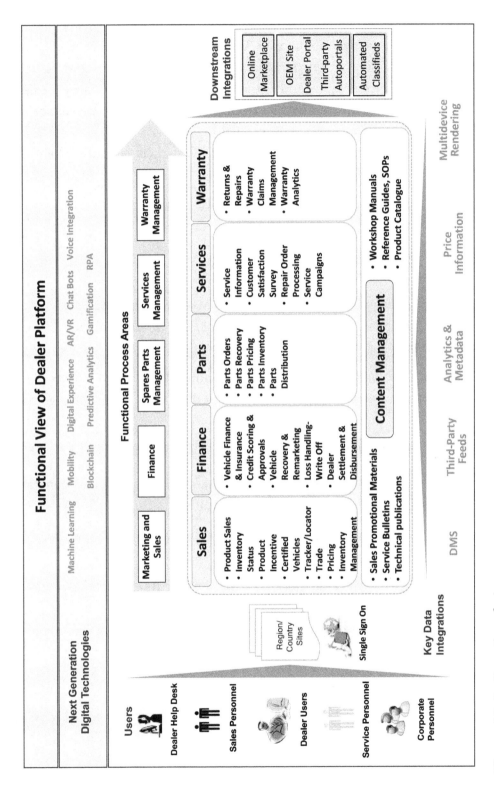

Figure 11-4. Functional view

The next-generation dealer platform needs various forward-looking features such as AI, chatbot, predictive analytics, mobility, AR/VR, blockchain, and such.

The platform needs multiple functional modules for managing parts (orders, pricing, inventory), sales (data, inventory, incentive, tracker), finance (vehicle finance, dealer settlement and such), warranty (returns and repairs, claims), and services (service information, customer satisfaction survey and such).

The content management capability of the DXP is used to implement the following features:

- *Sales promotion materials*: The sales team can create promotion content and target it for specific personas.

- *Bulletins and publications*: Share up-to-date vehicle-specific service bulletins and other publications with dealers.

- *Product catalogue and reference guides*: OEMs can seamlessly communicate sales promotions, product info, etc. through content standardized (or) customized across dealerships.

- *Technical information*: To train dealer personnel on technical aspects of the platform allows for faster product roll-out.

Seamless and Optimized Business Process

In order to address the challenges with existing business processes, the next-generation dealer platform is optimized to provide following features:

- One-stop deal processing and single-click submission of multiple artifacts for dealers

- Real-time incentive inquiry from multiple devices (mobile, iPad, desktop)

- Real-time incentive validation on submission of deal to avoid postsubmission changes and reconciliation issues

- Real-time updates on the artifact status

A sample optimization of end-to-end deal processing flow is depicted in Figure 11-5.

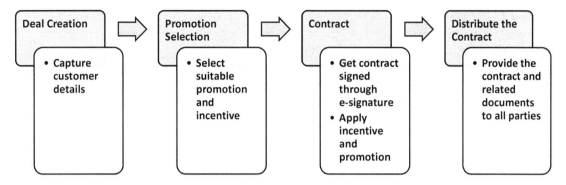

Figure 11-5. Deal processing flow optimization

Open-Source-Based Next-Generation Deal Digital Platform

The next-generation dealer digital platform is built on open-source technologies, and the guiding principles for the platform are listed in Table 11-5.

Table 11-5. Guiding Principles

Principles	Benefits
Leveraging open-source components	Leveraging open-source components like Angular JS, Bootstrap, Springboot, etc. Brings flexibility, agility, speed, and cost-effectiveness.
Micro ervices based	Event-driven architecture / domain-driven design Microservices are individually scalable and modular, providing high availability and performance.
API driven	Provide REST-based APIs that abstract the internal details of the application. The APIs form the contract for consuminig applications.
12-Factor apps	Supports the "software-as-a-service" principle for applications; this also brings scalability, resiliency, continuous delivery, maintainability, and information security

(*continued*)

Table 11-5. (*continued*)

Principles	Benefits
Scalable architecture for future cloud deployment	By keeping the microservices, microfrontend-based architecture can be easily moved to a cloud platform (platform as service model).
Keeping everything secured	Encryption in transit for all end-to-end traffic without exception; encryption at rest as per data classification
Automation enablement	Relentless automation in development and deployment process, and application and business processes Think of API as a service, not manual repetitive work.
Quality of Service (QoS) requirements from the beginning	Engineering culture to include usability, performance, scalability, releasability, and supportability as first-class concerns instead of an afterthought

The logical architecture of the dealer digital platform is shown in Figure 11-6.

CHAPTER 11　END TO END DXP CASE STUDY

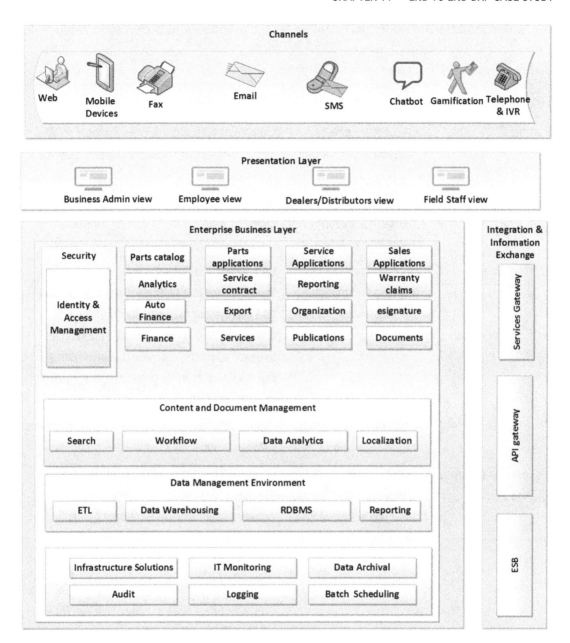

Figure 11-6. *Dealer platform logical architecture*

Chapter 11　End to End DXP Case Study

Various solution components mapping for the requirements are given in Table 11-6.

Table 11-6. *Solution Component Mapping*

Logical Component	Description	Shortlisted Component
Security	Security aspects of solution	Lightweight Directory Access Protocol (LDAP), Security Assertion Markup Language (SAML) components
Search	Enterprise content, navigation search capabilities	Apache Solr, ElasticSearch for domain data search
Microservices	Tools, frameworks to develop and manage microservices	SpringBoot and Docker Provides excellent features in rapid development and deployment of microservices. When desired to move to the cloud, the integrations to easily adopt microservice design patterns like intelligent routing, circuit breaking, config management, client-side load balancing, etc.
Content management	Solution for managing enterprise content	Drupal CMS-preferred considering the headless CMS and robust features in CMS products
Self-service business intelligence	Tool for creating, sharing simple/complex reports, both predefined, ad hoc, and self-served	Jaspersoft and Pentaho are open-source offerings. Jaspersoft is ideal for simple reporting. Pentaho has BI reporting capabilities.
UI/UX	User experience tool/frameworks	Custom using Angular JS, because of the flexibility it provides and ease of changing frameworks/libraries in this fast-changing UI/UX technology landscape
API Gateway	Gateway that would be the face of all channels via which integration services are provided	Kong, Tyk, DreamFactory

(*continued*)

Table 11-6. (*continued*)

Logical Component	Description	Shortlisted Component
DevOps	Tools/frameworks for DevOps	Maven / Jenkins
Analytics	Analytics solution framework	Google Web Analytics
Source code	For maintaining source code	Git, SVN, BitBucket
Functional testing	Testing the code for functional features	Junit
Performance testing	Conduct performance test	JMeter

Key solution components are detailed as follows:

UI Development

The next-gen dealer digital platform requires an intuitive UI that provides an engaging experience to users. Because of this requirement, we suggest developing responsive single-page application (SPA) following responsive web design (RWD) techniques using HTML5, Angular, Bootstrap, CSS, and Angular Material Design. The key design principle is to separate data from representation. SPA invokes APIs to get the required content and renders so that it is responsive. Features like internationalization/localization are implemented using frameworks like i18Next.

Domain Specific Microservices

As the platform will be transformed over time, it is important to keep delivering business value continuously. We can leverage microservices to be decomposed based on domains and subdomains. This will allow business functions to be independently developed and perhaps even coexist with other legacy functions before they are transformed.

Each microservice will have its own isolated stack include the data source. Docker containerization will be used to ensure portability across environments and platforms. For Java services, Spring Cloud libraries will be leveraged for implementation of various microservices patterns of service discovery, client-side load balancer, token relay, circuit breaker, and intelligent routing. Wherever available, the platform services will be given preference to language-specific libraries. Depending on the needs of the specific microservice, an appropriate datastore would be selected from relational, NoSQL, or Object Cache offerings of the platform for that microservice.

Drupal Headless CMS

A headless CMS is a back-end only CMS built from the ground up as a content repository that makes content accessible via a RESTful API for display on any device. Headless CMS functionalities provided by CMS solution providers such as Drupal deliver content via API, and these solutions also can be hosted on the cloud. Rich UI application will leverage these APIs to provide the content management functionalities in a next-gen dealer system. The APIs will be accessed via API gateway for content retrieval. Authentication and authorization will be done using security mechanisms like SSO, SAML, OAuth, access tokens, etc.

Innovations and Next-Generation Technologies in Dealer Platform

Given in Table 11-7 are innovations and usage of cutting edge digital technology that can be used for a dealer platform.

CHAPTER 11 END TO END DXP CASE STUDY

Table 11-7. Next-Generation Technologies for Dealer Platform

Technology	Description	Applicable Scenarios in Dealer Digital Platform
Chatbot	Using a chatbot for dealer and customer interaction	1. Enquiring inventory of car models and availability 2. Asking about the process flow steps for complex business process such as parts ordering
Mobility	Enable all key pages and functionality on smart phones and tablets.	1. Functions such as dealer dashboards, vehicle enquiry 2. Provide configurable home pages for dealers and sales team
Gamification	Applying game mechanics and game design techniques to engage and motivate all stakeholders to achieve their goals	1. Create dealer dashboard and sales dashboard to show the scoring and ranking to motivate dealers and sales team. 2. Provide real-time view of performance KPIs and metrics.
Digital experience	Provide integrated experience for all stakeholders.	1. Provide single-stop-shop experience for end to end deal management including pricing, application and deal closure. 2. Automate the key business processes.
Predictive analytics	Analytics for improving customer experience by identifying improvement opportunities at every touchpoint in the sales and service life cycle	1. Use predictive analytics to proactively forecast parts inventory based on sales. 2. Analyze the SLAs of the vehicle service process and warranty/claims process and identify the key bottleneck areas.
Automated integrated systems and dashboards	Seamless and streamlined integrated data from various business functions such as parts ordering, servicing, warranty	1. Provide dealer dashboard providing leads, warranty, customer satisfaction score. 2. Provide sales dashboard showing the region-wise sales and charts for the same.

(*continued*)

Table 11-7. (*continued*)

Technology	Description	Applicable Scenarios in Dealer Digital Platform
Artificial intelligence	Leverage AI to augment the chatbot and use for automate repetitive jobs, and use AI to gather insights from existing data.	1. Analyze warranty and claims data to identify the region-wise sales and leads. 2. Enable voice based queries for dealers 3. Analyze service rating and customer complaints to understand the issues with business processes.
Blockchain	Use of blockchain technology for ensuring transparency in information across stakeholders	1. Leverage blockchain to track the vehicle parts across OEMs, dealers, and customers. 2. Use blockchain technologies to get the complete vehicle ownership and service history.
Augmented/virtual/ mixed reality	Workforce of the future solution leveraging AR , VR, and MR	1. Using AR / MR for remote technical assistance 2. Use the AR and VR for dealer training.

Chapter Summary

- A vehicle dealer management B2B scenario is used in this case study. The key drivers and requirements for the dealer platform are next-generation user experience, business process improvements, improved information management, improved customer satisfaction, and a scalable and robust platform.

- In order to architect the next-generation dealer platform we need to do the pain point analysis, define solution tenets, devise solution design principles, and develop persona-based information architecture.

- The overall solution architecture of the next-generation vehicle dealer platform should define the functional view, create seamless and optimized business processes, leverage open source technologies, identify solution components mapped to key functional components, and use innovations

APPENDIX A

Open-Source Tools and Frameworks

A DXP uses open-source tools and frameworks to build its layers and mainly focuses on open-source technology to provide a digital ecosystem to develop invocative solutions and business processes. This section explains different frameworks that can be used in developing a DXP.

HTTP Accelerator

An HTTP accelerator is designed for content-heavy dynamic web application as well as APIs.

- Varnish (https://varnish-cache.org/)

A DXP uses Varnish to accelerate web content delivery to the client.

Web Server

A web server serves contents to the World Wide Web. A web server processes incoming network requests over HTTP and several other related protocols.

- Microsoft IIS (https://www.iis.net/)
- Nginx (https://www.nginx.com/)
- Apache web server (https://httpd.apache.org/)

A DXP uses these web servers to deploy their static contents, like HTML, CSS, scripts, images, etc. You can customize and use one of them to deploy and host their static content.

APPENDIX A OPEN-SOURCE TOOLS AND FRAMEWORKS

CSS Framework

A CSS framework is a framework that is meant to allow for easier, more standards-compliant web design using the Cascading Style Sheets language. CSS frameworks contain a grid structure for responsive web design.

- Bootstrap (https://getbootstrap.com/)
- Foundation (https://foundation.zurb.com/)
- Bulma (https://bulma.io/)
- Material UI (https://material-ui.com/)
- Semantic UI (https://semantic-ui.com/)

A DXP uses these open-source CSS frameworks to build UI designs. You need to check the compatibility of these frameworks with the scripting framework.

Scripting Framework

The scripting framework is a JavaScript framework. This library offers features that allow you to implement complex requirements:

- Angular (https://angular.io/)
- React (https://reactjs.org/)
- React Native (https://facebook.github.io/react-native/)
- NativeScript (https://www.nativescript.org/)
- Electron (https://electronjs.org/)

A DXP uses these open-source CSS frameworks to develop UI components along with CSS frameworks. You can choose permutations and combinations of CSS and scripting framework to develop UI components.

User Interface Management

UI management tools like a package manger, module bundler, task runner, or testing framework would help you to manage the modules and submodules in complex UI application. A module bundler puts all its dependency in one JS file. A task runner executes tasks based on the specific criteria to automate the UI build process.

- *Package manager*:
  - NPM (https://www.npmjs.com/)
  - YARN (https://yarnpkg.com/en/)
  - Bower (https://bower.io/)
- *Module bundler*:
  - Webpack (https://webpack.js.org/)
  - Browserify (http://browserify.org/)
  - Rollup (https://rollupjs.org/guide/en/)
- *Task runner*:
  - Grunt (https://gruntjs.com/)
  - Gulp (https://gulpjs.com/)
- *Testing*:
  - Mocha (https://mochajs.org/)
  - Jest (https://jestjs.io/)
  - Jasmine (https://jasmine.github.io/)
  - Cucumber (https://cucumber.io/)
  - Karma (https://karma-runner.github.io/latest/index.html)

A DXP uses a package manager and module bundler to integrate and manage the dependency package used while developing UI components, and uses a task runner to build tasks to minify and watch the changes while developing UI components. A DXP uses previously mentioned open-source testing frameworks for test-driven development and behavior-driven development.

APPENDIX A   OPEN-SOURCE TOOLS AND FRAMEWORKS

Integration

Integration provides a model for interaction and communication between mutually interacting software applications in service-oriented architecture (SOA).

- *Enterprise system bus (ESB)*:
    - Apache Camel (http://camel.apache.org/what-is-camel.html)
    - JBoss ESB (http://jbossesb.jboss.org/)
    - Open ESB (https://www.open-esb.net/)
    - Apache ServiceMix (http://servicemix.apache.org/)
- *Integration framework*:
    - Apache CXF (http://cxf.apache.org/)
    - Spring Integration (https://spring.io/projects/spring-integration)
    - Node Red (https://nodered.org/)
- *API Gateway*:
    - Gravitee (https://gravitee.io/)
    - Apiumbrella (https://apiumbrella.io/)
    - Apiman by RedHat (http://www.apiman.io)

The DXP uses and recommends open-source ESB architecture to develop the integration layer, but you can use another open-source integration framework to develop your DXP's applications and its integration layer. You can also use open-source API gateways to manage, authenticate, and scale the API integration layer. The ESB framework can handle large and complex integration, whereas an integration framework would be used in small- and medium-scale integrations.

Application Server

An application server is a component-based framework that resides in the middle tier of a server-centric architecture. It provides middleware services. The application is hosted on the application server.

APPENDIX A   OPEN-SOURCE TOOLS AND FRAMEWORKS

- Tomcat (http://tomcat.apache.org/)
- JBoss (http://jbossas.jboss.org/downloads)

A DXP uses open-source application servers to deploy the integration and middleware applications.

Server-Level Cache

Sever-level caches are standards-based caches that boost performance and simplify scalability. They are often used to speed up dynamic web applications by caching data and objects in RAM to reduce the number of times an external data source, like a database or API, is called.

- Jgroups (http://www.jgroups.org/)
- Ehcache (http://www.ehcache.org/)
- Memcached (https://memcached.org/)
- Redis (https://redis.io/)

A DXP uses open-source server level caches while developing an application to cache data objects.

Content Management Systems

A content management system (CMS) is a software application that is integrated with the DXP's applications to create and manage digital content.

- OpenCms (http://www.opencms.org/en/)
- TYPO3 (https://typo3.org/)
- Joomla (https://www.joomla.org/)
- Drupal (https://www.drupal.org/)

A DXP uses open-source CMS to provide a seamless digital experience that reach one's audience across multiple channels.

APPENDIX A OPEN-SOURCE TOOLS AND FRAMEWORKS

CMIS

Content Management Interoperability Services (CMIS) is an open standard that allows different content management systems to integrate with a DXP and control document management systems and repositories using web protocols.

- Apache Chemistry (https://chemistry.apache.org/)

A DXP uses open-source CMIS specification to query CMS and integrated CMS with multiple channels.

SQL Database

DXP uses Open source Database to store relational data.

- MySQL (https://www.mysql.com/)
- PostgreSQL (https://www.postgresql.org/)
- Maria DB (https://mariadb.org/)

NoSQL Database

A DXP uses an open-source NoSQL database to store nonrelational data. These databases have capabilities to push JSON to one's application in real-time.

- MongoDB (https://www.mongodb.com/)
- Redis (https://redis.io/)
- CouchDB (http://couchdb.apache.org/)
- Cassandra (http://cassandra.apache.org/)
- RethinkDB (https://www.rethinkdb.com/)

IoT Framework

IoT frameworks and platforms are also called IoT middleware; the purpose is to function as a mediator and integrator between the hardware and application layers. Primary tasks include data collection from the devices over different protocols and networks.

- Eclipse Kura (https://www.eclipse.org/kura/)
- Node-RED (https://nodered.org/)
- Flogo (https://www.flogo.io/)
- Iotivity (https://iotivity.org/)
- AllJoyn (https://openconnectivity.org/developer/reference-implementation/alljoyn)

A DXP uses open-source IoT middleware frameworks to connect, control, and integrate multiple devices with the DXP's applications.

Distributed Data Streaming

A distributed streaming platform simplifies data integration between a DXP's systems. A stream is a pipeline to which one's applications receive data in real time.

- Apache Kafka (https://kafka.apache.org/)
- Apache ActiveMQ (http://activemq.apache.org/)
- Redis (https://redis.io/)

A DXP uses an open-source distributed streaming framework to provide real-time data access using producer, consumer, and broker streams.

Analytics Engine

Spark is packaged with higher-level libraries and includes support for SQL queries, streaming data, machine learning, and graph processing.

- Apache Spark (https://spark.apache.org/)

DXP uses an open-source analytics framework to analyze huge and diverse data sources.

APPENDIX A OPEN-SOURCE TOOLS AND FRAMEWORKS

Distributed Processing

Apache Hadoop is a collection of open-source software utilities that facilitate using a network of many computers to solve problems involving massive amounts of data and computation.

- Apache Hadoop (https://hadoop.apache.org/)

A DXP uses a distributed processing framework that process large datasets across a cluster of computers.

Machine Learning Library and Framework

Machine learning is a part of artificial intelligence (AI) that provides a DXP the ability to automatically learn and improve. In the following are mentioned open-source framework and library implement machine learning algorithms, categorized as supervised or unsupervised. These libraries easily get integrated with an application that can access data and use it for learning of its own.

- Tensorflow (https://www.tensorflow.org/)
- PyTorch (https://pytorch.org/)
- Scikit-learn (http://scikit-learn.org/)
- Deeplearning4j (https://deeplearning4j.org/)
- Apache Ignite (https://ignite.apache.org/)
- Apache Mahout (https://mahout.apache.org/)
- Apache SINGA (https://singa.incubator.apache.org/en/index.html)

A DXP uses an open-source machine learning and deep learning library and framework to automate different tasks on different DXP layers, using diversified programming languages.

Blockchain Frameworks

Blockchain is an upcoming technology; frameworks listed in the following provide enterprise business blockchain capabilities where DXP assets can be tracked and audited with enabling distributed ledger capabilities.

- Hyperledger (https://www.hyperledger.org/)
- MultiChain (https://www.multichain.com/)
- HydraChain (https://github.com/HydraChain/hydrachain)
- Corda (https://www.corda.net/)
- Openchain (https://www.openchain.org/)
- IOTA (https://www.iota.org/)

A DXP uses an open-source blockchain framework and platform to deploy and execute smart contracts and build the DXP's applications on blockchain infrastructure.

Augmented and Virtual Reality

Augmented reality (AR) and virtual reality (VR) are increasingly used in technology. It's reality created by the use of technology to add digital information on an image of something. AR is used in apps for smart phones and tablets. AR applications use one's mobile or tablet camera to show one a view of the real world and augment information, including text and images, on top of that view.

- ARToolkit (https://github.com/artoolkit)
- ARKit (https://developer.apple.com/arkit/)
- ARCore (https://developers.google.com/ar/)
- AR.js (https://aframe.io/blog/arjs/)

A DXP uses open-source AR and VR frameworks to build AR-VR applications on top of the DXP's integrated environment.

APPENDIX A   OPEN-SOURCE TOOLS AND FRAMEWORKS

Enterprise Search Engine

An enterprise search engine would be integrated with a DXP to assist in locating important information within a short period of time.

- *Solr stack*:
  - Apache Solr (http://lucene.apache.org/solr/)
  - Banana (https://github.com/lucidworks/banana)
- *Elastic stack*:
  - Elasticsearch (https://www.elastic.co/)
  - Logstash (https://www.elastic.co/products/logstash)
  - Beats (https://www.elastic.co/products/beats)
  - Kibana (https://www.elastic.co/products/kibana)

A DXP uses an open-source enterprise search engine platform to integrate it with the DXP's ecosystem and provide search engine capabilities to one's application.

Containerization

Containerization is a lightweight option to provide full machine virtualization that involves encapsulating an application in a container with its own operating system and environment. This provides many benefits of loading an application onto a virtual machine (VM) hence the application can be run on any suitable physical machine without any worries about dependencies.

- Docker (https://www.docker.com/)

A DXP uses an open-source container platform to build, manage, and secure one's application and wrap in a container.

APPENDIX A   OPEN-SOURCE TOOLS AND FRAMEWORKS

Containerization Orchestration

Containerization orchestration is a system for automating deployment, scaling, and management of containerized applications.

- Docker Swarm (https://docs.docker.com/engine/swarm/)
- Kubernetes (https://kubernetes.io/)

A DXP uses containerization orchestration systems for automating deployment, scaling, and management of containerized applications.

Source Code Management

SCM is a software versioning and revision control system. Software developers use SCM to maintain current and historical versions of files such as source code, web pages, and documentation.

- Git (https://git-scm.com/)
- Apache Subversion (https://subversion.apache.org/)

A DXP uses an open-source SCM tool to manage and control versioning of an application developed.

Continuous Integration and Continuous Delivery

Continuous integration (CI) and continuous delivery (CD) are used to automate all sorts of tasks related to building, testing, and delivering or deploying software.

- Jenkins (https://jenkins.io/)
- Gerrit (https://git.eclipse.org/r/Documentation/install.html)

A DXP uses an open-source DevOps platform to build, deploy, and review the DXP's applications.

APPENDIX B

Sample Code

This sample source code will demonstrate the BXP application's two UI components (also called widget or portlet), which are Account and Transaction. You will understand the development of the UI layer and integrations layer, with mocking the data services using mocking frameworks, as shown in Figure B-1. You can include a database while doing integration. Mock services can be replaced with actual services after development of the DXP's application. Sample source code is available for download from the book's GitHub repository. You can access it at www.apress.com/9781484243022 and click the Source Code button.

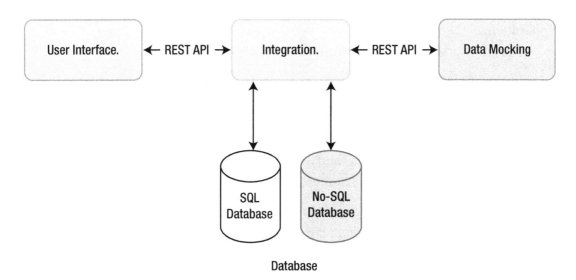

Figure B-1. Developing UI and Integration

APPENDIX B   SAMPLE CODE

User Interface

You can use any one of the following UI scripting library or UI scripting frameworks to develop the UI of the application along with its supporting CSS framework:

- Angular
- Native Script
- React Native
- React
- Electron
- Vue

We have chosen the Angular framework for this example because it's based on MVC architecture, whereas React's JavaScript library just helps us to update the view of the user. But you can integrate FLUX to control the workflow of React-based UI application. You can choose one of the previously listed frameworks or libraries depending upon your requirements, design and implementation of which will be suitable for your application.

Points to evaluate before choosing a UI technology stack are:

- *MVC*: MVC pattern has three components: model, view, and controller.
  - Model is bound with view as well as controller.
  - View is a user interface that binds the model with the Document Object Model (DOM) and display data to the user and also enables the user to modify the model.
  - Controllers are responsible for controlling the flow of the application; if you make a web services request the controller is responsible for providing a response back to the application.

  The MVC pattern provides control over business logic implemented on UI scripting; hence you should consider whether your application needs the MVC pattern.

- *Data binding*: Data binding would help you to establish a relationship between your business logic with the DOM. Angular has two-way binding, whereas React has one-way binding.

- *Rendering*: You need to consider which type of data rendering your application requires; it is preferred to use server-side rendering frameworks because of less load time as compared with client-side rendering, and would optimize your application for web crawling.

- *Performance*: Two-way data binding would impact the performance of the complex application, while one-way binding would not impact as much.

You need to look at your requirement and the aforementioned considerations together to evaluate and decide on a framework.

Integration

We have chosen Apache Camel as the integration framework because it provides easy implementation of integration of a variety of different applications, which use several protocols, frameworks, and technologies. It is a lightweight mini enterprise service bus (ESB) framework that implements all enterprise integration patterns.

You can use one of the following open-source frameworks to develop your integration layer:

- Apache Camel
- Apache ServiceMix
- WSO2 ESB
- Open Source Mule ESB
- Open Source Talend ESB

You can use open-source mini-ESB frameworks if you want to integrate two to three protocols and technologies, for example:

- Reading the files
- Soap to REST conversions
- XML to JSON conversions
- Reading or writing to data streams

Otherwise you can also use commercial ESB products for large-scale integration projects.

APPENDIX B   SAMPLE CODE

Data Mocking

We choose Swagger to mock web services because it gives features to design, build, test, and share API along with mocking the web services as per the recommendation of OpenAPI guidelines, which are meant to provide a standard format to unify how an industry defines and describes RESTful APIs. You can use any of the following API mocking frameworks:

- Swagger
- RAML
- WireMock

Implementation and Logic

The Angular technology stack (Typescripts, Webpack, NPM, and NodeJS), as shown in Figure B-2, would contain UI implementation and application logic to handle the state of the client-side application. Client-side applications have web hooks; these hooks are responsible for getting the data from the server-side application. Complex business logic and calculation should be avoided on the client side, as one can manipulate it on the browser's developer tools.

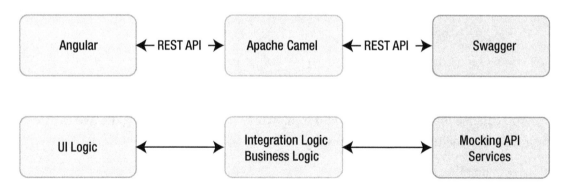

Figure B-2. Implementation and logic

The Apache Camel framework, as shown in Figure B-2, has a responsibility to provide API endpoints while the client-side Angular technology stack strictly handles the user interface. Complex business logic (like calculation of taxes, deduction of taxes, etc.) is implemented along with integration at the server-side. The server-side application contains a business controller, which is responsible for business logic, and integration controllers that are responsible for sending and receiving data to other application using Apache Camel's messaging components and endpoints.

Swagger will mock the third-party web services, which will be removed while integrating the DXP's application with third-party systems.

Deployment

We have chosen Apache Tomcat to deploy our application. You can deploy Angular code on the Apache web server and the integration application on Apache Tomcat.

Development

In development you can serve your Angular code through Angular CLI. Apache Camel resides in a Spring container and this application is deployed on a Tomcat instance, as shown in Figure B-3.

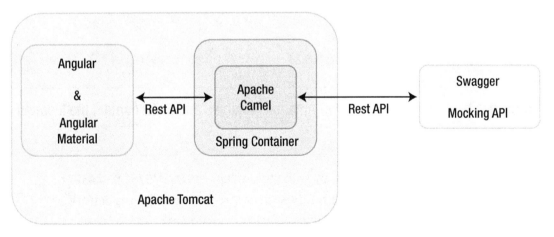

Figure B-3. *Deployment*

APPENDIX B SAMPLE CODE

Production

In production you can deploy the code on a cluster of Apache web server and Apache Tomcat server instances. You can do load balancing of your application through the Apache web server.

Prerequisite

Install the following tools and software:

1. Intellij IDEA community edition – Version: Latest (https://www.jetbrains.com/idea/download/#section=windows)

2. Visual Studio Code – Version: Latest (https://code.visualstudio.com/)

3. Java Development Kit (JDK) – Version: 8 and Java Runtime environment (JRE) – Version: 1.8.0_191. (https://www.oracle.com/technetwork/java/javase/downloads/index.html)

4. NodeJS – Version: v8.12.0 and NPM – Version: 6.4.1 (https://nodejs.org/en/)

5. Git: Version Control System – Version: Latest (https://git-scm.com/)

6. Apache Maven – Version: 3.3+ (https://maven.apache.org/download.cgi)

7. Apache Tomcat – Version: 8.5.34 (https://tomcat.apache.org/)

Important Make sure that environmental variables and PATH contain the location of JDK, JRE, NODE, and GIT.

You can access the source code by going to www.apress.com/9781484243022
Then clone the source code and folder structure from the Git repository https://github.com/apress/building-digital-experience-platforms.

API Specification and API Mocking

Let's start with developing and mocking API using Swagger UI. An API consumer of the application can use this specification to develop their solution by mocking the services till development phase. This specification is the contract between the API consumer and provider, which helps them to integrate the application in a production environment. You can download the Swagger designer from the following locations. We have used OpenAPI-Specification Version 2.

- https://swagger.io/
- https://github.com/swagger-api
- https://github.com/OAI/OpenAPI-Specification/blob/master/versions/2.0.md

Swagger-UI

You can access the API specification by using Swagger-UI.

> *Step 0*: Open the Mocking_Services\swagger-ui folder in the source code.
>
> Clone Swagger UI: https://github.com/swagger-api/swagger-ui
>
> *Step 1*: Clone Swagger UI using the following command:
>
> git clone https://github.com/swagger-api/swagger-ui.git
>
> *Step 2*: Run "**npm install**"; you need an open Internet connection because this command will copy the dependency from the NPM repository to your local machine.
>
> *Step 3*: Run "**npm start**"; when the server starts, it will be running on port number 3002. Access the following URL:
>
> http://localhost:3002

APPENDIX B   SAMPLE CODE

Swagger-Editor

You can change the API specification by editing the API specification file in the Swagger editor.

Step 1: Run "**npm install -g http-server**". NPM will install the http-server modules in the local repository.

Step 2: Run swagger-editor by running the following command:

http-server swagger-editor

Step 3: It will run on port Number 8080, or check the log of the preceding command and access the URL, as shown in Figure B-4.

http://localhost:8080/

Figure B-4. *OpenAPI specification for transaction and account*

Step 4: Click File ➤ Open and Example File and select abcBanking.yaml as shown in Figure B-5.

340

APPENDIX B SAMPLE CODE

Figure B-5. abcBanking.yaml

Step 5: abcBanking.yaml contains API specification for account and transaction. One can created a mock server by selecting mock server for abcBanking.yaml as shown in Figure B-6.

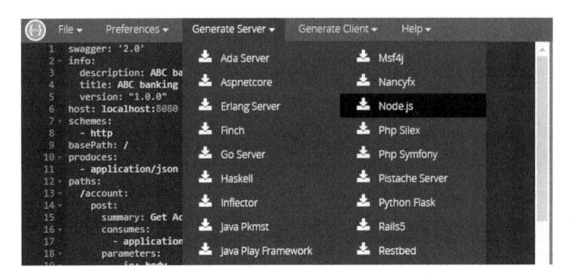

Figure B-6. Node server generation

Step 6: Generate server by selecting the Node.js option; it will create a zip file. Unzip (**nodejs-server-server-generated**) and open the command prompt at the same location.

341

APPENDIX B   SAMPLE CODE

Swagger-Server

Continue with the following steps:

> *Step 7*: Run "**npm start**" inside the unzipped folder. It will download the dependency and the server will start at port number 8080, or check the logs of command "**npm start**" as shown in Figure B-7.
>
> http://localhost:8080/docs

Figure B-7. *API server*

> *Step 8*: Click default; you can access account and transaction services. You can access the services at the following locations using REST Client, as these are POST method requests.
>
> - Account: http://localhost:8080/account
>
> - Transaction: http://localhost:8080/transaction

UI Screen Mocking on Node-RED

You can use Node-RED for modeling your application using APIs and UI screens. Node-RED is a programming tool for wiring together hardware devices, APIs, web services, and quickly making a model to visualize the data flow and UI screens. It provides a browser-based editor that makes it easy to wire together flows using the wide range of nodes in the palette, which can be deployed to its runtime in a single-click. You access Node-RED documentation from the following URLs:

342

APPENDIX B   SAMPLE CODE

- https://nodered.org/
- https://github.com/node-red/node-red

    Step 1: Open folder location from the source code \Intergartion_Framework\ABC_Bank_Integration\Node-Red_Based_Integration

    Step 2: Clone Node-RED using the following command:

    git clone https://github.com/node-red/node-red.git

    Step 3: Run "**npm install**"; you need an open Internet connection, because this command will copy the dependency from NPM repository to your local machine.

    Step 4: Run "**npm start**"; the server starts on port number 1880, as shown in Figure B-8, or access the log of the previous commands and access the following URL: http://127.0.0.1:1880/

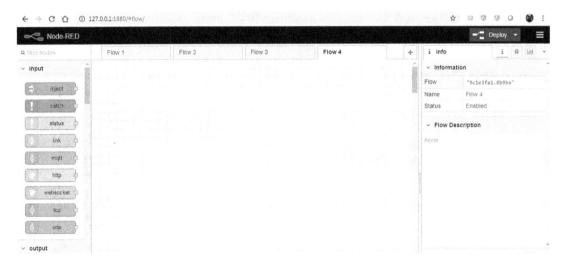

Figure B-8. *Node-RED*

    Step 5: Click (Menu ➤ Manage palette ➤ Install).

    Step 6: Search and install (node-red-dashboard) palette as shown in Figure B-9.

APPENDIX B   SAMPLE CODE

Figure B-9. *Install dashboard palette*

Step 7: Open folder (Intergartion_Framework\ABC_Bank_Integration\
Node-Red_Based_Integration\node-red-flow.txt) and copy the content
of the (node-red-flow.txt) file, as shown in Figure B-10.

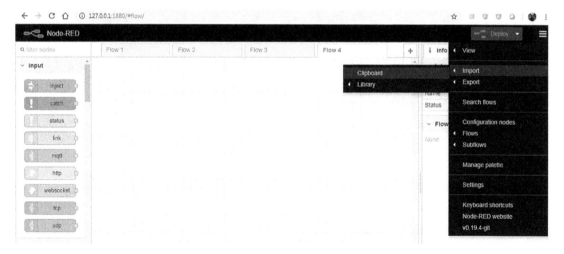

Figure B-10. *Import flow*

APPENDIX B   SAMPLE CODE

Step 8: Click *(Menu ➤ Import ➤ Clipboard)* and paste the content of the *(node-red-flow.txt)* file. Click the import button, as shown in Figure B-11.

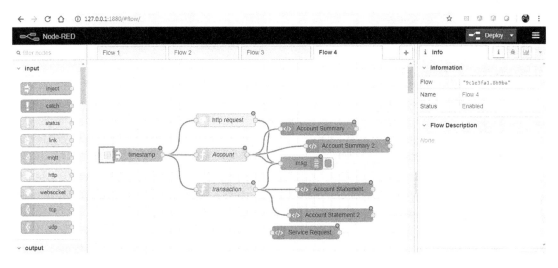

Figure B-11. *Account and transaction flow*

Step 9: Click the maroon deploy button on the right corner. After clicking the deploy button, click and **inject timestamp** to the flow by clicking the left side timestamp input highlighted in the red color box.

Step 10: Access the dashboard from the following URL, as shown in Figure B-12: http://127.0.0.1:1880/ui/

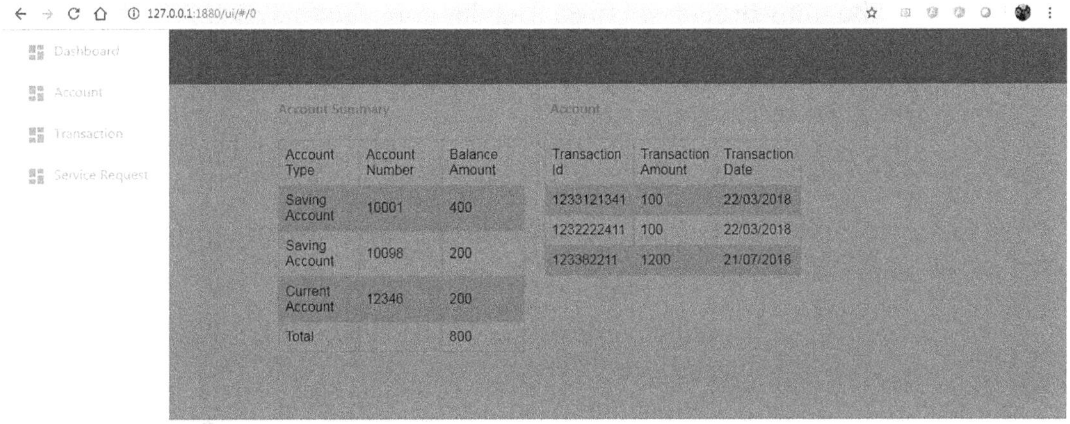

Figure B-12. *Account and transaction mockup dashboard*

345

APPENDIX B SAMPLE CODE

> **Note** You can install the Node-RED library for integration like MySQL, MongoDB, Swagger, etc. in a Flow-based wiring tool from the following location: `https://flows.nodered.org/`

Apache Camel

Apache Camel (`http://camel.apache.org/`) provides lot of useful components that support many libraries and frameworks such as Hibernate, Apache Spark, Apache-CXF, Apache Kafka, Restlet, Servlet, FTP, etc. These framework helps in integrating data between two different systems. For example, using the Camel Servlet or Camel Restlet, you can pull data from web services, transform it, and send it to another system or front-end application over REST API calls.

Prerequisite step:

- Swagger mocking API server should be running before moving forward.

In this application, we go over an integration example:

- Reading the account and transaction data from the Swagger API specification
- You can calculate the total balance in the controller, which fetches data from mock API and exposes another transformed API.
- Send the transformed API to an Angular application over REST API calls.

Here are the full details of the example:

- Read the data from Swagger API endpoints.
- Access the API content and perform the transformation using a custom processor.
- Create the Camel routes to expose the new API endpoint.

APPENDIX B   SAMPLE CODE

Build Automation System

Automation Systems will help you to provide support and maintenance of multiproject builds that are expected to be quite huge, and helps you to maintain the dependencies of the application with regard to third-party modules and parts, the build order, as well as the needed plug-in. It will download libraries and plug-ins from the different repositories and then put them all in a cache on your local machine. It also allows for incrementally adding to your build, because it knows which parts of your project are getting updated.

To demonstrate the application, we have chosen Maven. But you can also use other build automation tools such as Gradle.

Add Dependency

This integration application is a Maven-based Project's; hence, add the following dependencies in the POM.xml file of the project. It is available in the project's root folder for your reference.

- *Camel-core*: The main dependency for Apache Camel
- *Camel-spring*: Enables us to use Camel with Spring
- *Camel-stream*: An optional dependency, which you can use to display some messages on the console while routes are running
- *Spring-context*: The standard spring dependency, required in our case as we are going to run Camel routes in a Spring context (Spring container)
- *Spring-core*: The main dependency of Spring
- *log4j*: Log4j is a Java-based logging utility that enables us to implement logging in our application.
- *Jetty*: The Jetty plug-in is used to as a lightweight server to instantly run the application and test the development work.

APPENDIX B ■ SAMPLE CODE

Figure B-13 provides an overview of Apache Camel-based integration deployed on Tomcat. It will help you to see that the rest of the URL is constructed from different layers of the application.

- Application server host and port: `http://localhost:8001/`
- Application WAR filename: integration.0.0.1
- The URL pattern setting from web.xml: camel
- End point URI of the Camel servlet endpoint:
  - account
  - transaction

Hence the endpoint URI for account is: `http://localhost:8001/integration.0.0.1/camel/account`

And for transaction, it is: `http://localhost:8001/integration.0.0.1/camel/transaction`

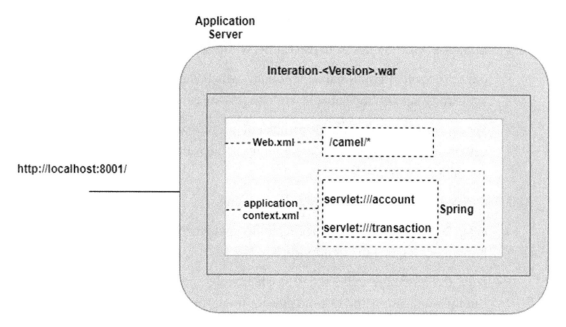

Figure B-13. *Camel servlet integration application deployed in application server*

Camel Servlet Component

The Camel Servlet component is used to process incoming HTTP requests, where the HTTP endpoint is bound to a published servlet. The servlet component is implemented by the following servlet class:

org.apache.camel.component.servlet.CamelHttpTransportServlet

To create a Camel servlet endpoint in a Camel route, define a servlet endpoint URI as the following syntax, as shown in Figure B-14.

servlet://RelativePath[?Options]

Web.xml File

To deploy the Apache Camel integration application, you must provide a properly configured web.xml file. In the integration project, the web.xml file (Listing B-1) is stored at the following location: Integration/src/main/webapp/WEB-INF.

Listing B-1. Web.xml File for the Integration application Example

```xml
<?xml version="1.0" encoding="ISO-8859-1"?>
<web-app version="2.4" xmlns="http://java.sun.com/xml/ns/j2ee"
xmlns:xsi="http://www.w3.org/2001/XMLSchema-instance"
xsi:schemaLocation="http://java.sun.com/xml/ns/j2ee http://java.sun.com/
xml/ns/j2ee/web-app_2_4.xsd">
<!-- location of spring xml files -->
<context-param>
<param-name>contextConfigLocation</param-name>
<param-value>classpath:applicationContext.xml</param-value>
</context-param>

<!-- the listener that kick-starts Spring -->
<listener>
<listener-class>org.springframework.web.context.ContextLoaderListener
</listener-class>
</listener>
```

APPENDIX B SAMPLE CODE

```xml
<!-- Camel servlet -->
<servlet>
<servlet-name>CamelServlet</servlet-name>
<servlet-class>org.apache.camel.component.servlet.
CamelHttpTransportServlet</servlet-class>
<load-on-startup>1</load-on-startup>
</servlet>

<!-- Camel servlet mapping -->
<servlet-mapping>
<servlet-name>CamelServlet</servlet-name>
<url-pattern>/camel/*</url-pattern>
</servlet-mapping>

</web-app>
```

listener/listener-class: This element launches the Spring container, as shown in Figure B-14.

context-param: This element specifies the location of the Spring XML file, camel-config.xml, in the WAR. The Spring container will read this parameter and load the specified Spring XML file, which contains the definition of the Camel route.

servlet/servlet-class: Specifies the org.apache.camel.component.servlet.CamelHttpTransportServlet class, which implements the Camel Servlet component servlet-mapping/url-pattern. It determines which URLs are routed to this servlet. In general, the servlet URL has the following form: http://Host:Port/WARFileName/URLPattern.

Where the base URL, http://Host:Port, is determined by the configuration of the web server, the WARFileName is the root of the WARFileName.war WAR file, and the URLPattern is specified by the contents of the url-pattern element.

Assuming that the application server port is set to 8001, the integration.0.0.1application would match URLs of the following form: http://localhost:8001/integration.0.0.1/camel/*.

The Camel route for this example, defined in a Spring XML file in applicationContext.xml, using Camel's XML DSL syntax, is shown in Listing B-2.

Listing B-2. Route Definition for the Camel Servlet Example

```
<camelContextxmlns="http://camel.apache.org/schema/spring">

<route>
<!-- incoming requests from the servlet is routed -->
        <from uri="servlet:///test"/>
                <transform>
        <simple>Version 1.</simple>
                </transform>
</route>

<route>
<!-- incoming requests from the account servlet is routed -->
        <from uri="servlet:///account"/>
        <process ref="accountControllerRequest"/>
        <camel:touri="direct:/v1/account" />
        <process ref="accountControllerResponse"/>
</route>

<route>
<!-- incoming requests from the transaction servlet is routed-->
        <from uri="servlet:///transaction"/>
        <process ref="transactionControllerRequest"/>
        <camel:touri="direct:/v1/transaction" />
        <process ref="transactionControllerResponse"/>
</route>

<camel:routestreamCache="true">
        <!-- outgoing requests from the account to http services
        account -->

        <camel:fromuri="direct:/v1/account" />
        <camel:removeHeaderspattern="CamelHttp*"
excludePattern="CamelHttpMethod" />
        <camel:setHeaderheaderName="HttpMethod">
        <camel:constant>POST</camel:constant>
        </camel:setHeader>
```

APPENDIX B   SAMPLE CODE

```xml
            <camel:setHeaderheaderName="CamelHttpMethod">
            <camel:constant>POST</camel:constant>
            </camel:setHeader>
            <camel:setHeaderheaderName="Content-Type">
            <camel:constant>application/json</camel:constant>
            </camel:setHeader>
            <camel:setHeaderheaderName="CamelHttpQuery">
            <camel:simple></camel:simple>
            </camel:setHeader>
    <camel:setBody>
                                            <camel:simple>{"accountNumber":
                                            "12345"}</camel:simple>
            </camel:setBody>
            <!--Swagger Mocking service http services account URL -->
            <camel:touri="http://localhost:8080/account" />
</camel:route>

        <camel:routestreamCache="true">
        <!-- outgoing requests from the transaction to http services
        transaction -->
            <camel:fromuri="direct:/v1/transaction" />
            <camel:removeHeaderspattern="CamelHttp*"
excludePattern="CamelHttpMethod" />

            <camel:setHeaderheaderName="HttpMethod">
            <camel:constant>POST</camel:constant>
            </camel:setHeader>
            <camel:setHeaderheaderName="CamelHttpMethod">
            <camel:constant>POST</camel:constant>
            </camel:setHeader>
            <camel:setHeaderheaderName="Content-Type">
            <camel:constant>application/json</camel:constant>
            </camel:setHeader>
            <camel:setHeaderheaderName="CamelHttpQuery">
            <camel:simple></camel:simple>
            </camel:setHeader>
```

```xml
            <camel:setBody>
            <camel:simple>{
            "userID": "12345",
            "firstName": "Sourabh",
            "lastName": "Sethi"
            }</camel:simple>
            </camel:setBody>
<!--Swagger Mocking service http services transaction URL -->
            <camel:touri="http://localhost:8080/transaction" />
</camel:route>
            </camelContext>
```

The servlet URL, `servlet: ///account`, specifies the relative path, `/account`. The complete URL to access this servlet is the following, as shown in Figure B-14: `http://localhost:8001/integration.0.0.1/camel/account`.

- *accountControllerResponse*: This processor class is responsible for transforming and converting responses from account services.

- *accountControllerRequest*: This processor class is responsible for transforming request headers and body before heading to account services.

The servlet URL, `servlet: ///transaction`, specifies the relative path, `/transaction`. The complete URL to access this servlet is the following, as shown in Figure B-14: `http://localhost:8001/integration.0.0.1/camel/transaction`

- *transactionControllerResponse*: This processor class is responsible for transforming and converting responses from transaction services.

- *transactionControllerRequest*: This processor class is responsible for transforming request headers and body before heading to transaction services.

APPENDIX B SAMPLE CODE

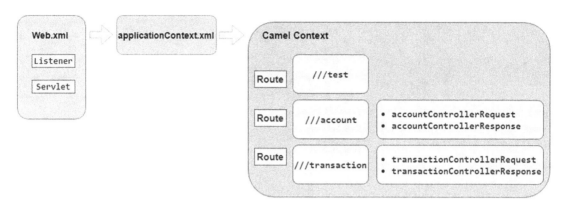

Figure B-14. Overview of Camel application

Run the Integration Application

To run the integration application's sample code, please follow the following steps.

1. Ensure that the Swagger mock API server is working as shown in the API Specification and API Mocking section.

2. Navigate to the following location in the sample source code.

   Integration_Framework\ABC_Bank_Integration\Apache_Camel_Based_Integration\integration

3. Ensure that the latest Maven version is installed on your machine by executing the following command:

   mvn -v

4. Execute the following command to test and run the application on the local Jetty server:

   mvn jetty:run -Djetty.port=8001

5. The following services will be up and running:
   - http://localhost:8001/integration/camel/account
   - http://localhost:8001/integration/camel/transaction

6. In the production environment, one can deploy an application WAR file, available at location \integration\target\integration-0.0.1.war in the application server.

7. You can open this application in Intellij IDEA IDE.

Angular

Let's start with developing a front-end responsive angular application for mobile as well as desktop. We have created integration applications till now. We will use an API provided by the integration application to an Angular application so that data can be populated using this web API. You can read the Angular documentation from the following locations:

- https://angular.io/
- https://angular.io/docs

    Step 1: Navigate to source code location (Angular\ABC Bank\angular-material)

    Step 2: Execute the "**npm install**" command; it will download dependency modules from remote repository to local repository.

    Step 3: Check the (proxy.conf.json) file to implement revere proxy for integrating web services while doing development, as shown in Figure B-15. In production you can deploy these angular production files directly into your back-end application or by serving it via Apache 2 or Nginx.

```
{
    "/integration/*": {
      "target": "http://localhost:8001",
      "secure": false,
      "logLevel": "debug",
      "changeOrigin": true
    }
}
```

APPENDIX B   SAMPLE CODE

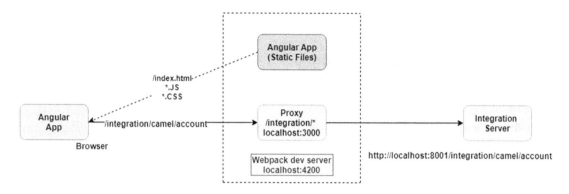

Figure B-15. *Proxy web services for developing Angular application*

Step 4: Run "**npm start**"; it will start the application, as shown in Figure B-16, at the following endpoint:

http://localhost:4200/first-page

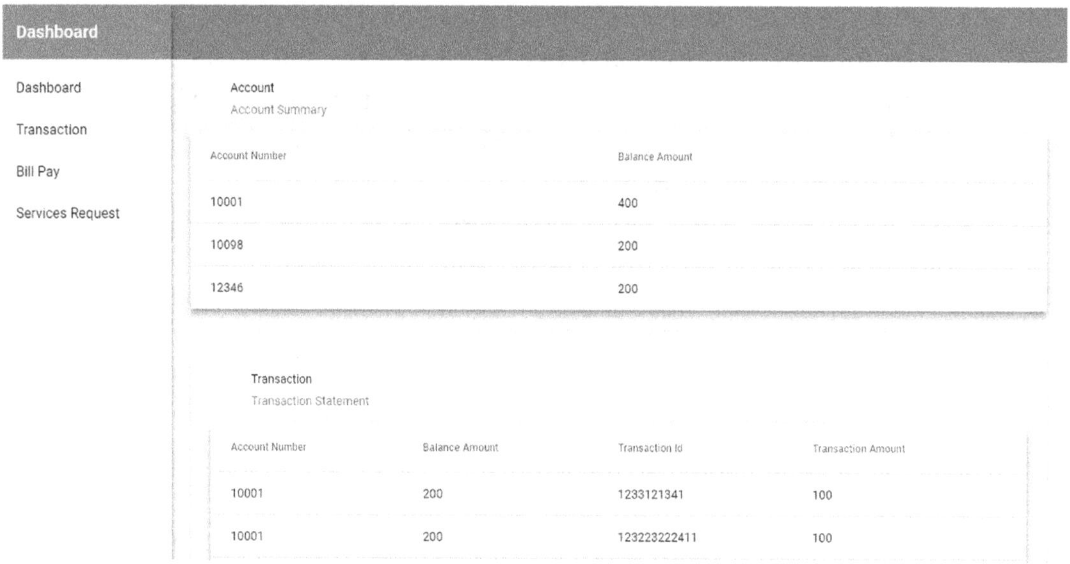

Figure B-16. *ABC banking dashboard*

You can modify the component or extend the components in the project structure shown in Figure B-17. You can open this project in VS code IDE.

356

APPENDIX B   SAMPLE CODE

Figure B-17. *ABC banking angular project structure*

Microservices Architecture

You can replace the Apache framework (monolithic architecture) with microservices architecture. Microservices architecture is an architectural style that structures an application as a collection of loosely coupled services that implement business capabilities. Microservices architecture is an alternative pattern that addresses the limitations of monolithic architecture. You can deploy a service as a (Docker) container image and deploy each service instance as a container; you can cluster Docker using Docker clustering frameworks such as:

- Kubernetes (https://kubernetes.io/)
- Marathon (https://mesosphere.github.io/marathon/)
- Docker Swarm (https://docs.docker.com/engine/swarm/)

ABC banking microservices uses Spring Boot and Eureka Server. You can start building a Spring microservices project by selecting the required modules and generating the project from the following location: https://start.spring.io/

357

APPENDIX B SAMPLE CODE

Microservices Components

Microservices components are comprised of Config Server, Eureka Discovery server, and services components such as account and transaction services, and gateway services, as shown in Figure B-18. The user sends a request to integrator services (API gateway); the API gateway is responsible for redirecting the request to microservices instances.

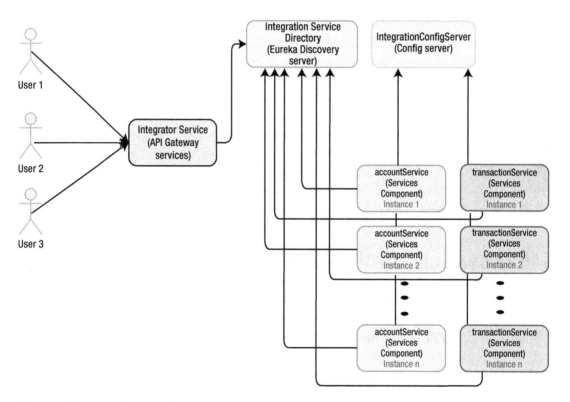

Figure B-18. *Microservices components*

Config Server: This server helps you to keep the properties file centralized and shared by all microservices and manage all the microservices properties files; those files are version controlled using Git. One thing to remember is that every microservice communicates with Config Server to get properties values.

To run the ABC banking Config Server, take the following steps:

    Step 1: Navigate to the source code.

    \Intergartion_Framework\ABC_Bank_Integration\Microservices\integrationConfigServer

    Step 2: Execute "**mvnspring-boot:run**" to start the services.

    Step 3: Hit the following URL to check the configurations:

    http://localhost:9090/config/default

    The response will be like

    {"name":"config","profiles":["default"],"label":null,"version":null,"state":null,"propertySources":[]}

Eureka Discovery server: Microservices is decentralization of the different components based on the business features. It can be scaled as per need, so for particular microservices, there can be multiple instances. These services are deployed as a container and these containers have dynamic IP addresses, so to track all instances of a service, a manager service will be needed. If other services need to communicate with each other, it contacts a discovery service to get the instance of another service.

To run the ABC banking Eureka Server, use the following steps.

    Step 1: Navigate to the source code.

    \Integration_Framework\ABC_Bank_Integration\Microservices\integrationServiceDirectory

    Step 2: Execute "**mvnspring-boot:run**" to start the services.

    Step 3: Hit the following URL to check the configurations, as shown in Figure B-19:

    http://localhost:9091/

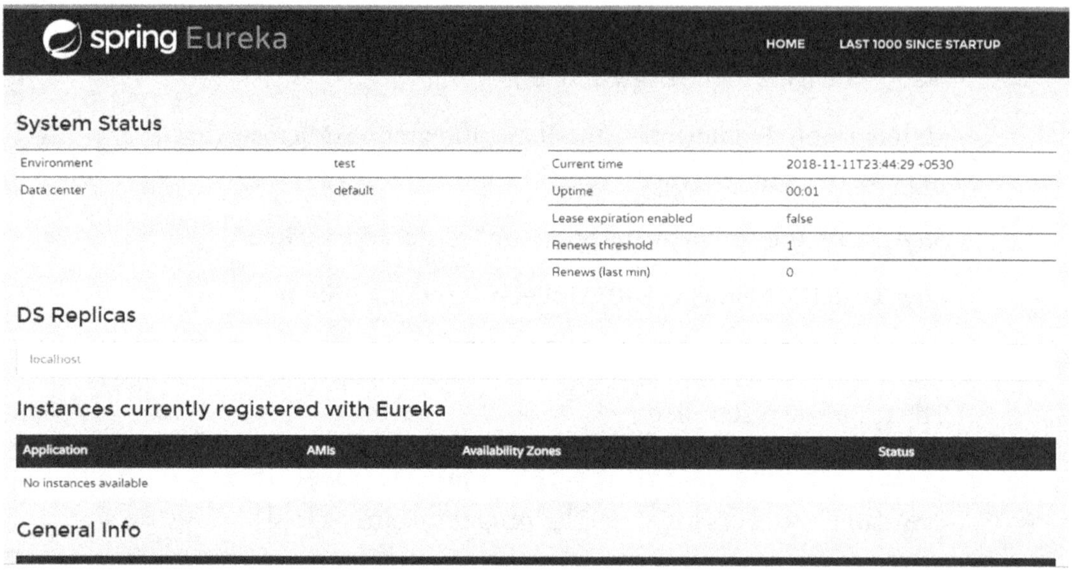

APPENDIX B SAMPLE CODE

Figure B-19. Eureka server

Services components: The goal of microservices is to break down complete business functionality into several independent small features that will communicate with each other. It provides modular architecture with proper encapsulation and properly defined boundaries.

To run the ABC banking account services, use the following steps:

Step 1: Navigate to the source code.

D:\DXP\Code\Intergartion_Framework\ABC_Bank_Integration\Microservices\accountService)

Step 2: Execute "mvnspring-boot:run" to start the services.

Step 3: Hit the following URL to check the configurations:

http://localhost:8083/account/findall

The response will be:

{"list":[{"accountNumber":10001,"balanceAmount":"200"},
{"accountNumber":10002,"balanceAmount":"200"},
{"accountNumber":10003,"balanceAmount":"200"}]}

Note that static data is configured, but you can write your own logic in AccountContoller and get the data from the database by integrating the database using a JPA module.

To run the ABC banking transaction services, take the following steps:

Step 1: Navigate to the source code.

Intergartion_Framework\ABC_Bank_Integration\Microservices\transactionService

Step 2: Execute "mvnspring-boot:run" to start the services.

Step 3: Hit the following URL to check the configurations:

http://localhost:8082/transaction/findall

The response will be:

{"list":[{"accountNumber":10001,"transactionId":"800",
"transactionAmount":"300","balanceAmount":"200"},
{"accountNumber":10002,"transactionId":"800",
"transactionAmount":"300","balanceAmount":"200"},
{"accountNumber":10003,"transactionId":"800",
"transactionAmount":"300","balanceAmount":"200"}]}

Note the following:

- Static data is configured, but you can write you own logic in TransactionController and get the data from the database by integrating the database using JPA module.

- You can replace microservices with Apache Camel services by integrating the endpoints in an Angular application.

- You can open these microservices projects in Intellij IDEA IDE.

- You can access the Eureka Server URL and can check the services and instance registered with Eureka Server as shown in Figure B-20.

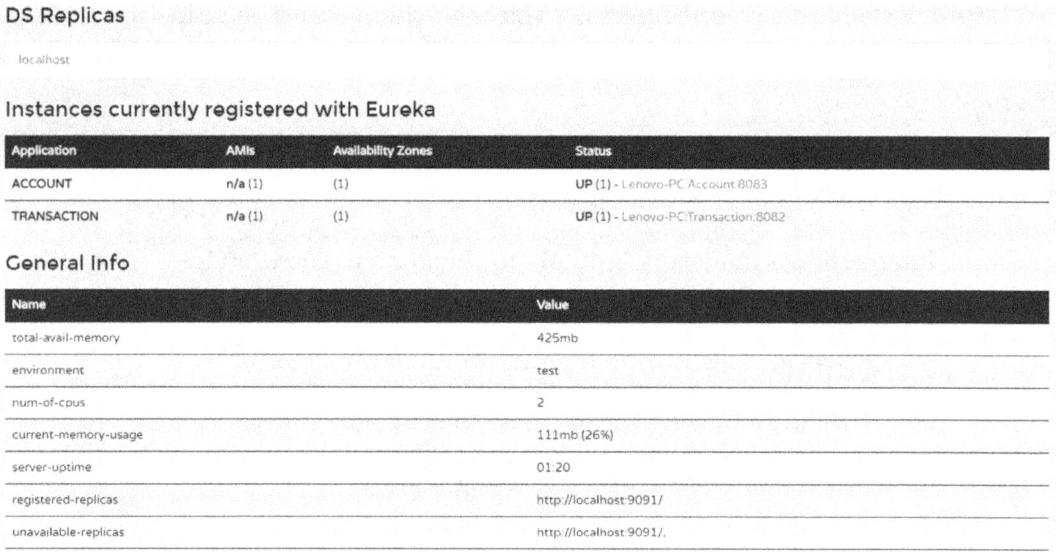

APPENDIX B SAMPLE CODE

DS Replicas

localhost

Instances currently registered with Eureka

Application	AMIs	Availability Zones	Status
ACCOUNT	n/a (1)	(1)	UP (1) - Lenovo-PC:Account:8083
TRANSACTION	n/a (1)	(1)	UP (1) - Lenovo-PC:Transaction:8082

General Info

Name	Value
total-avail-memory	425mb
environment	test
num-of-cpus	2
current-memory-usage	111mb (26%)
server-uptime	01:20
registered-replicas	http://localhost:9091/
unavailable-replicas	http://localhost:9091/,

Figure B-20. *Account and transaction microservices*

Gateway service: Because every microservice publishes a REST API, it is hard to manage so many endpoint URLs. If you want to build an authentication and authorization framework, which ought to be implemented across all the microservices, a gateway service will help, which is Internet facing. The client will call only one endpoint and it delegates the call to an actual microservice and all the authentication or security checking will be done in the gateway service.

Step 1: To Run the service integrator, navigate to the following source code location:

\Intergartion_Framework\ABC_Bank_Integration\Microservices\integratorService

Step 2: Execute "**mvnspring-boot:run**" to start the services.

Step 3: Hit the following URLs to check the configurations:

- Account: `http://localhost:8081/integrator/feign/account`
- Transaction: `http://localhost:8081/integrator/feign/transaction`

Step 4: You can open these microservices projects in Intellij IDEA IDE.

APPENDIX B   SAMPLE CODE

You can use Zuul to load balance microservices, as shown in Figure B-21. When Zuul receives a request, it routes the request to one of the physical locations available. The process of caching the location of the service instance and forwarding the request to the actual location is provided out of the box.

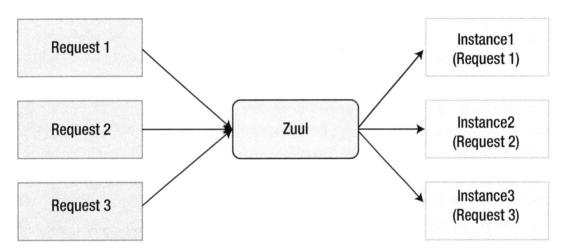

Figure B-21. *ZUUL load balancing*

Docker

After checking that all the services are running locally, you can use Docker to containerize these microservices.

You can download and install Docker CE on your local machine from the following URL: `https://docs.docker.com/get-started/`.

Components

Dockerfile: It is a text file that contains all the instruction to build a Docker image;, these files contain steps to copy files and do installation. For more information, you can check the following link: `https://docs.docker.com/engine/reference/builder/`

Docker Composer: Docker Composer creates and spawns multiple containers. It helps to build the required environment.

APPENDIX B SAMPLE CODE

You can create an individual container for each service. The following is a list of containers for this example:

- integrationConfigServer
- integrationServiceDiretory
- accountService
- transactionService
- integratorService

Summary

- We have gone through mocking third party RESTful API using OpenAPI specification using Swagger.
- We have gained an understanding of the application by prototyping it using Node-RED.
- We have gotten hands-on experience on integrating the Apache Camel framework (back-end integration) with an Angular application (front-end application).
- We understood and implemented microservices architecture using Spring Boot.

APPENDIX C

Further Reading

Koelsch, George. *Requirements Writing for System Engineering.* Apress, 2016.

Shivakumar, Shailesh Kumar. *Architecting High Performing, Scalable and Available Enterprise Web Applications.* Morgan Kaufmann, 2014.

Shivakumar, Shailesh Kumar. *Complete Guide to Digital Project Management.* Apress, 2018.

Shivakumar, S. K. *A Complete Guide to Portals and User Experience Platforms.* Chapman and Hall/CRC, 2015.

Shivakumar, S. K. (n.d.). "DevOps for Digital Enterprises." Infosys. Retrieved October 20, 2018, from https://www.infosys.com/digital/insights/Documents/devops-digital-enterprises.pdf

Shivakumar, S. K. (n.d.). "Digital Experience Platforms – An Overview." Infosys. Retrieved October 20, 2018, from https://www.infosys.com/digital/insights/Documents/digital-experience-platforms.pdf

Shivakumar, S. K. *Enterprise Content and Search Management for Building Digital Platforms.* John Wiley & Sons, 2016.

Shivakumar, Shailesh Kumar. "Method and system for presenting personalized content." U.S. Patent Application No. 14/489,410.

Shivakumar, S. K. (n.d.). "Reimagining online experiences with digital experience platforms." Retrieved October 20, 2018, from https://www.infosys.com/digital/insights/Documents/reimagining-digital-experience-platforms.pdf.

Sourabh Sethi. "Healthcare Blockchain leads to Transform Healthcare Industry." *International Journal of Advance Research, Ideas and Innovations in Technology* 4.1, 2018. https://www.ijariit.com/manuscripts/v4i1/V4I1-1359.pdf.

Index

A

Accessibility, 216
AccountControllerRequest, 353
AccountControllerResponse, 353
Aggregator pattern architectures, 80–81
AI automation design
    building model, 106
    Chatbots, 107–108
    data preprocessing, 106
    goals, 106
    improving and tuning model, 107
    RPA, 106
    testing model, 107
    training model, 107
Amazon web services (AWS), 227
Analytic services, 271
Angular framework, 134
Angular library
    bootstrap, 135
    Gulp, 136
    material UI, 135
    NativeScript, 136
    swagger, 135
    Webpack, 136
Angular technology stack (ATS), 75, 133
Apache Camel
    angular application, 355–357
    run integration application, 354
Apache Lucene, 110
API ecosystem (integration), 64, 67
APIs and microservices, 268
Applicability to bank landscape, 267
Application server, 224, 324–325
Application-specific security analysis, 195
Artificial intelligence (AI), 121, 262, 268
Asset tracking, 95
Auditing and logging, 209
Augmented reality (AR), 269, 329
Augmented-Virtual Reality (AVR)
    integration service layer, 112
    presentation layer, 111
Authentication and authorization, 208
Automation, 279
Availability, 215, 225

B

Back office services, 273
Banking experience platform (BXP)
    application consistency, 141
    components, 145–147
    dashboard, 142–144
    key tenets, 16–17
    KPIs, 24–25
    layouts/containers, 142
    location consistency, 141
    requirements, 18–21
    technical capabilities, 21–24
    three Ps, 21

INDEX

Big data and NoSQL
    components, 102
    containerization, 105
    database, 104
    data streams and processing, 103
    employee engagement, 106
    ETL, 102
    IoT model-based algorithm, 105
    open-source projects, 102
    search and query web services, 103
    SQL *vs.* NoSQL, 105
    train and test predictive
        data model, 102–103
Biometric authentication, 268
Blockchain, 96, 268
    architecture, 100
    claim management, 101
    components
        communication layer, 99
        consensus layer, 99
        crypto layer, 99
        data services and API
           managements, 99
        data store layer, 99
        identify services, 99
        smart contract layer, 99
    definition, 96
    digital identity, 101
    distributed ledger, 97
    EHR, 100
    enterprise network, 98
    KYC, 100
    letter of credit, 101
    libraries, 97
    loyalty and rewards, 101
    network, 97–98
    platforms, 98
    PoE, 101
    public network, 98
    smart contracts, 97
    transactions, 97
    types, 98
Blog, 92
Bootstrap, 135
Brand value, 62
Business intelligence (BI), 275
Business layer
    authorization, 85
    business logging, 85
    business rules validation, 85
    data access layer, 85–86
    model binding, 85
    redirection, 84
Business to business (B2B)
    architectures, 81

C

Camel-core, 347
Camel Servlet, 349, 351–353
Camel-spring, 347
Camel-stream, 347
Cascading style sheets (CSS), 124
Centralized access, 201
Chatbots, 92
    AI strategy, 107
    bot engine, 107
    integration, legacy system, 108
    open-source bot frameworks, 108
    presentation layer, 107
Cloud testing, 211–212
Code backup, 230
Cold backup option, 231
Compartmentalization, 201

INDEX

Component architecture, 130–132
Configurability, 216
Containerization, 330
Content backup, 230
Content delivery network (CDN), 222
Content management, 267
Content management interoperability services (CMIS), 326
Content management system (CMS), 70, 325
Content services, 273
Context-param, 350
Continuous delivery (CD), 331
Continuous deployment (CD), 114
Continuous integration (CI), 114, 331
Continuous learning and improvement, 63, 67
Create, read, update and delete (CRUD), 210
Cross-site request forgery (CSRF), 56
Cross-site scripting (XSS) filters, 57
Crowd-based P2P lending, 269
Cryptocurrency, 268
CSS framework, 322
Customer insights gathering, 262

D

Data access object (DAO), 85–87
Data access policy definition, 201
Data backup, 230
Database level security, 210
Database server, 225
Data binding, 334
Data design, 67
Data layer, 86–87
Data mocking, 336
Data sharing, 210
Data standardization, 164
Data validation, 189
Dealer platform
    functional view, 310, 312
    next-generation technologies, 319–320
    open-source technologies, 313–314, 316
    pain point analysis, 299–301
    sample optimization, 312–313
    solution design principles, 304–307
    solution tenets, 302–304
    user centricity, 308–309
Defense in depth, 201
    backup and synch jobs, 203
    DR, 203
    firewall and proxies, 202
    monitoring infrastructure, 202
    server hardware level protection, 202
Denial of service (DoS), 202, 226
Deployment
    development, 337
    production, 337–338
    tools and software, 338
Design
    brand value, 62
    IA, 62
    interaction design, 62
    layers, 70–72
    principles, 63
    visual, 62
Development life cycle, 129–130
DevOps, 68
    CI and CD, 114
    containerization, 113–114
Digital asset management (DAM), 8
Digital banking
    business process-related trends, 269
    technology related trends, 268–269

369

INDEX

Digital experience platform (DXP), 3
    boundaryless banking, 4
    business drivers, 14–16
    case study, 254–256
    challenges and solutions, 13
    core components, 7–11
    dealer platform, case study
        (*see* Dealer platform)
    evolution, 11–12
    experience requirements, 32
    functional requirements (*see* Functional
        specification document (FSD))
    key tenets, 5
    mobility requirements, 32, 41–43
    nonfunctional requirements, 32
    reference architecture, 5
    requirement elicitation and elaboration
        methods, 28–31
    security framework (*see* Security
        framework)
    security requirements, 32
Digital facelift for user experience, 279
Digital-first bank, 269
Digital imperatives, 25
Digital maturity assessment
    analytics, 286
    business agility, 285
    business process, 282
    collaboration and social media
        interaction, 284
    data management, 286
    governance, 283
    infrastructure, 286
    IT alignment, 285
    leadership, 283
    organization culture, 281
    user engagement, 284

Digital open ecosystem, 262
Digital opportunities, 287
Digital platform strategy, 65
    API ecosystem (integration), 67
    continuous learning and
        improvement, 67
    data design, 67
    DevOps, 68
    infrastructure design, 67
    principle, 66
    touch points, 67
Digital transformation
    design phase, 287
    execution phase, 288
    road map, 288
    tools and methods, 289
        artificial intelligence and machine
            learning, 293
        big data, 293
        content management, 292
        DevOps, 291
        digital experience capabilities, 292
        integration model, 290
        security, 293
        social and collaboration, 290
        user experience, 289
        web analytics, 291
Digital wallets, 96, 269
Digitization, 280
Digitizing banking business models, 279
Digitizing existing banking systems, 278
Disaster recovery (DR), 203, 215
    activities, 230
    As-Is system analysis, 232
    implementation, 230–231
    maintenance, 231
    planning, 229–230, 232–233

scope and objective, 232
strategy, 228–229
requirements
    RPO, 57–58
    RTO, 57
Distributed denial of service (DDoS), 202, 226
Docker, 363
Docker composer, 363
Dockerfile, 363
Document object model (DOM), 257, 334
Domain specific microservices, 318
DR site
    setup, 230
    switching, 230
DXP solution
    cloud deployment model, 226–228
    cloud hosting
        integration design, 226
        platform coexistence, 225–226
        security, 226
        tiered architecture, 224–225
    technology stacks, 132–133

E

ECMAscript, 130
Efficiency, 216
Elastic Stack
    Beats, 109
    Elastic Cloud, 110
    Elasticsearch, 110
    Kibana, 110
    Logstash, 109
Electronic healthcare records (EHR), 100
Enable new integrations, 264
End-user experience, 235
Enterprise integration, 266
Enterprise resource planning (ERP), 150

Enterprise search engine
    elastic stack, 109–110
    indexing, 109
    matching, 109
    processing, 109
    query processing, 109
    SolrStack, 110–111
    sources, 109
Enterprise service bus (ESB), 335
Error handling, 209
Experience requirements
    dashboard user story, 39–40
    language criteria, 38
    navigational routers, 39
    supported browsers, 38
    supported device, 37
Extensibility, 216
Extract, transform, and load (ETL), 102

F

Fast delivery (infrastructure), 64
File management, 209
File storage server, 225
Firewall, 224
Flexibility, 216
Flexible integration middleware
    API gateways, 167
    EAI *vs.* SOA *vs.* ESB *vs.* Microservices, 165–166
    MOU, 167
Forums, 92
Front office services, 271
Functionality, 267
Functional specification
        document (FSD), 32
    account use case, 33–34
    transaction use case, 35–36

G

Gamification, 268
Google calendar API, 92

H

Horizontal services, 273
Hot backup option, 231
HTTP accelerator, 321

I

Infrastructure design, 67–68
Implementation and logic, UI, 336–337
Information architecture (IA), 62
Information management, 188–189
Information security policies, 201
    access controls, 205
    archival and retention, 206
    auditing and logging, 206
    availability, 206
    classification, 204
    confidentiality, 207
    definition, 205
    description, 203
    destruction, 206
    incident response, 207
    integrity, 207
    ownership identification, 204
    private data, 207–208
    process creation, 203–204
    sharing, 206
    storage, 206
Infrastructure/capacity analysis, 250
Initial login performance, 255
Integrated analytics, 122
Integration, 335
    patterns, 162–164
    services, 9
    systems, 161–162
Integration layer
    advantages of microservices, 83
    architectures
        aggregator pattern, 80–81
        B2B, 81
        SOA, 81–82
    BXP, case study, 176–179
    channel patterns, 78
    components and patterns, 79
    consideration, 150–152
    data formats, 153–154
    data interoperability, 82
    data standardization, 164
    design, 83–84
    endpoint patterns, 79
    highly coupled, 78
    Kubernetes cluster, 83
    loosely coupled, 78
    management patterns, 79
    message components, 80
    message construct, 79
    message patterns, 80
    microservices and monolithic
        services, 82–83
    patterns, 162–164
    practices, 173–175
    protocols, 158
    routing patterns, 79
    security, 171–172
    services, 155–156
    styles, 157–158
    systems, 161–162
    technology, 168–170
    transformation patterns, 79

Interaction design, 62
International electrotechnical commission (IEC), 203
Internationalization, 122
International Organization for Standardization (ISO), 203
Internet of Things (IoT), 269
Interoperability, 216
Intuitive architecture, 120
Intuitive data, 64
IoT integration design, 94
    application layer, 94
    ARIoT, AllJoyn and Iotivity, 95
    asset tracking, 95
    banking through wearable, 96
    digital wallets, 96
    fast-growing technology, 93
    integration layer, 93
    ML libraries, 95
    open-source software components, 95
    physical sensing layer, 93
    sensors, 93
    smart cities and real-time streaming data, 95
    smart payment contract, 96

J

Jasmine, 136
Jetty, 347
JSON-RPC, 159
JSON web tokens (JWTs), 57
JSON-WSP, 160

K

Karma-Mocha-Chai, 136
Key performance indicators (KPIs), 120
Knowledge management (KM) portals, 92

Know your customer (KYC), 100
Kubernetes, 114

L

Layer performance optimizations
    database, 258
    presentation, 257
    server, 257–258
Lean portal services, 271
Least privilege by default, 201
Leveraging modern digital technologies, 262
Live chat, 91
Load balancer, 224
log4j, 347
Log analysis, 250
Low customization and high configuration, 266

M

Machine learning, 328
Maintainability, 216
Maintenance requirements
    monitoring, 51
    serviceability, 51
    SLA, 50
Microservices architectural integration technology, 170
Microservices architecture, 357–358
    banking transaction services, 361
    components, 358, 360
    config server, 358–359
    Eureka server, 359–362
    gateway service, 362
    JPA module, 361
    services components, 360
    ZUUL, 363

INDEX

Middleware layer
   application logging, 89
   application monitoring, 88
   auditing, 89
   components, 87–88
   server logging, 89
   server monitoring, 88
   transaction processing, 89
Mid-office services, 272
   campaign/marketing services, 272
   commerce services, 272
   search services, 272
   self-learning services, 272
   social and collaboration services, 272
Mobility, 265
Model-view-controller (MVC), 125, 271
Model-view-viewmodel (MVVM), 125
Modularity, 216
Monolithic architectural integration technology, 168–169
Mutual memorandum of understanding (MOU), 167
MVC pattern, 334

N

Natural language generation (NLG), 108
Natural language processing (NLP), 107
Natural language understanding (NLU), 108
Next-gen communication, 265
Next-generation digital bank
   attributes, 263–264
   DXP features, 265–267
Node package manager (NPM), 130, 137
Node-RED, 342–343

   account and transaction flow, 345
   import flow, 344
   mockup dashboard, 345
   palette, 344
Nonfunctional requirements (NFRs), 43, 245

O

Object relational mapping (ORM), 255
Open platform, 264
Open-source framework
   application server, 324–325
   blockchain, 329
   CSS, 322
   distributed streaming, 327
   integration, 324
   IoT, 327
   scripting, 322
Open-source technologies
   domain specific microservices, 318
   headless CMS, 318
   principles, 313–314, 316–317
   UI development, 317
Open-source tools
   Apache Hadoop, 328
   CD, 331
   CI, 331
   CMIS, 326
   CMS, 325
   HTTP accelerator, 321
   NoSQL database, 326
   SCM, 331
   SQL database, 326
   UI management, 323
   web server, 321
Optimized business models, 262

P

Page hits analysis, 48–49
Page response time (PRT), 244
Payment banks, 269
Payment card industry (PCI), 209, 265
Performance, 216, 335
Performance debugging framework
    common performance issues, 252–254
    component/system/layer, 251
    root cause, 247–250
    steps, 247
Performance optimization,
        web pages, 236–238
Performance requirements
    page hits analysis, 48–49
    page response time, 47–48
    use case, 46
Performance testing, DXP
    design, 240
    execution and reporting, 241–242
    metrics, 243–244
    requirement analysis, 239–240
    sprints, 239
    framework
        critical transactions, 245
        prediction, 247
        qualitative assessment, 245
        quantitative assessment, 246
        workload model, 245
Persona-based information
        architecture, 307–308
Personalization, 121
Platform design phases, 70
    delivery, 69
    design, 69
    explore and elaborate requirements, 69
    prototype, 69
    SDLC, 69
    validation, 69
Presentation component, 125–127
Presentation layer
    Chatbot, VR, AR, Alexa, and voice
        assistance, 73
    CSS and scripting framework, 74
    define, 73
    design, 73
    prototype, 73
    test, 73
    touch points, 72
    UI deployment, 76–77
    UI management, 75–76
    user experience, 72
Proof of existence (PoE), 101
Providing omnichannel capabilities, 279

Q

Quality attributes
    archival and retention, 221
    availability, 220
    logging and auditing, 221–222
    performance, 222
    reliability, 219
    scalability, 219
    security, 218
    usability, 217

R

Ramp-up test, 242–243, 250
React technology stack (RTS), 75
    flux, 138
    Jest, 138
    react native, 138
    semantic UI, 137

INDEX

Real-time data streaming, 95
Recovery point objective (RPO), 57, 220
Recovery time objective (RTO), 57–58, 220
Redux-MobX, 138
Reference functional architecture, 273–275
    analytics module, 275
    banking services module, 275
    BI, 275
    campaign module, 275–276
    content management module, 276
    data management module, 276
    loan module, 275
    open platform features, 276
    payment module, 275
    security services, 276
    technology transformation, 276–278
Reference technology architecture, 269
    back office services, 273
    delivery support, 273
    front office services, 271
    gamification, 273
    horizontal services, 273
    infrastructure and maintenance services, 273
    mid-office services, 272–273
    security and identity services, 273
Reimagining the banking experience, 278
Remote procedural call (RPC), 151
Rendering, 335
Robotics process automation (RPA), 106, 264
Robust platform, 266
Rollout protocols, 52–53

S

Sales channels, 287
Scalability, 44–46, 215, 225
Scripting framework, 322

Search engine optimization (SEO), 7, 58, 123
Secure incident management, 210
Secure sockets layer (SSL), 228
Security, 216, 267
    providers, 224
    testing, 211
Security and identity services, 273
Security awareness and training, 210
Security framework
    checklist, 196–199
    coding review, 193
    data validation, 189
    fund management, 196
    information management, 188–189
    key elements, 184
    key vulnerabilities, 190–192
    layers, 184–186
    password policies, 187–188
    service security management, 189
    session management, 188
    testing, 193–194
    transaction management, 195
    web security testing, 194
Security requirements
    authenticity and authorization, 55–56
    integrity, 56
    session management use case, 54–55
Self-learning and continuous improvement, 263–264
Server configuration analysis, 250
Server side performance issues, 255
Server-side rendering (SSR), 134
Service bus architectures (SOA), 81–82
Service enablement, 279
Service level agreements (SLAs), 215
Service security management, 189
Session management, 188

Sever-level caches, 325
Simple object access protocol
    (SOAP), 150, 158
Single-page applications (SPAs), 271
Single sign-on (SSO), 267, 293
Six-layered approach, 65
    API ecosystem (integration), 64
    continuous learning and
        improvement, 63
    fast delivery (infrastructure), 64
    intuitive data, 64
    touch points, 63
    user interface, 63
Sizing of DXP, 222–224
Smart cities, 95
Smart contracts, 97
Smart payment contract, 96
Social and collaboration design
    blogs, 92
    calendar, 92
    Chatbot, 92
    forums, 92
    KM portal, 92
    live chat, 91
    social interaction, 90
    social media API, 92
    software applications, 91
    tools, 89
    wiki, 92
Social and collaboration services, 9
Social media banking, 269
Social software applications, 91
Software as a service (SaaS), 5
Software development kits (SDKs), 265
Software development life cycle
    (SDLC), 69, 209

SolrStack, 110–111
Solution tenets, 302–304
Source code management (SCM), 331
Spring-context, 347
Spring-core, 347
SQL *vs.* NoSQL, 105
Stability, 216
Standardization and centralization, 279
Standard operating procedures (SOPs), 210
Supplier portal and distributor portal, 255
Supply chain management (SCM), 52
Support multiple sales channels, 263
Swagger-Editor, 340–341
Swagger-Server, 342
Swagger-UI, 339
Swarm, 114
Syntactically awesome style sheets
    (SASS), 135
SysAdmin, audit, network, security
    (SANS), 209

T

Thumb rules, web page, 236–238
Time to first byte (TTFB), 243
Touch points, 63, 67
TransactionControllerRequest, 353
TransactionControllerResponse, 353
Two-speed digital enablement, 279

U

UI deployment, 76–77
UI frameworks
    data flow, 139
    language, 139
    performance, 139

INDEX

UI management
    components and dependencies, 75
    CSS and scripting framework, 75
    module bundler, 75
    package manager, 75
    task runner, 75
    testing, 75
Unified and collaborative approach, 142
Usability, 216
User acceptance testing (UAT), 254
User-centric experience redesign, 262
User experience, 72, 267, 287
User experience enhancement services, 271
User interface (UI) design, 63, 334–335
    components
        hooks, 127
        layouts, 123–124
        MVC pattern, 125
        MVVM pattern, 125–126
        navigational router, 124
        pages, 123
    features
        dashboard, 120
        integrated analytics, 122
        internationalization, 122
        intuitive architecture, 120
        personalization, 121
        responsive interface, 120–121
        SEO, 123
    operable, 140
    perceivable, 140
    robust, 141

V

Varnish cache, 77
Virtual private cloud (VPC), 226
Virtual reality (VR), 269, 329
Visual design, 62, 127–129

W

Warm backup option, 231
Web and Http caches, 76
Web content accessibility guidelines (WCAG), 217
Webpack, 136
Web security testing, 194
Web server, 224
Web services
    JSON-RPC, 159
    JSON-WSP, 160
    SOAP, 158
    XML_RPC, 159
Web.xml File, 349–350
Widget/portlet, 333
Wiki, 92
Workflow and orchestration module, 272

X, Y

XML_RPC, 159

Z

Zed attack proxy (ZAP), 293

CPSIA information can be obtained
at www.ICGtesting.com
Printed in the USA
LVHW102349010519
616359LV00003B/50/P